ENGINEERING QUANTITIES AND SYSTEMS OF UNITS

ENGINEERING QUANTITIES
and
SYSTEMS OF UNITS

Rhys Lewis

B.Sc.Tech., C.Eng., M.I.E.E.
Llandaff College of Technology, Cardiff

HALSTED PRESS DIVISION
JOHN WILEY & SONS, INC.
NEW YORK

Published in the United States and Canada by
Halsted Press Division
John Wiley & Sons, Inc., New York

Library of Congress Cataloging in Publication Data

Lewis Rhys.
Engineering Quantities and Systems of Units

1. Weights and Measures. 2. Units. 3. International System of Units. I. Title.

QC88.L53   389   72–3115
ISBN 0–470–53377–3

WITH 38 ILLUSTRATIONS AND 9 TABLES

© 1972 APPLIED SCIENCE PUBLISHERS LTD

Printed in Great Britain by Galliard Limited, Great Yarmouth, England

Preface

At the present time the subject of measurement and the units in which quantities are measured are of prime importance. The move towards an internationally recognised system of units for general as well as technical use is accelerating for various reasons, not the least among them being increased economic cooperation between countries.

Meanwhile it is not sufficient to teach only the one system of units. Although the United Kingdom, for example, is in the process of 'metrication', it will be some time before the everyday user thinks as well as trades in International Units and part of this conversion is so often hampered by a basic lack of understanding of the systems now going out of use. In the United States, of course, the foot–pound–second systems will continue in use for some time and the need for a thorough understanding of gravitational units, for example, is still firmly established from a practical point of view.

This book covers all the important everyday quantities and the unit systems used to measure them. In addition, topics such as heat, light and electricity are considered in the units in which all countries now measure them, the International System (SI). A final chapter on the method of dimensions should prove most useful as, among other things, an aid to unit conversion.

From an educational viewpoint the book covers the main topics of the mechanics and electrical content of a considerable number of courses including the City and Guilds of London Institute and National Certificate courses in the United Kingdom and many high school technical–vocational courses and Technical Institute technology studies in the United States and Canada. In addition, it is hoped it will serve as a useful reference book for established engineers and technicians in industry and also for teachers and lecturers concerned with science and technology in schools and colleges.

Llandaff   RHYS LEWIS
South Wales
January 1972

Contents

CHAPTER ONE

Systems of Units

1.1 INTRODUCTION

The number of quantities, and, even more confusing, the number of different names given to what is apparently the same quantity, that must be understood by engineers and technicians is quite astonishingly large and, certainly to a newcomer, may be confusing in their multitude and apparent complexity. It is not always made apparent that all these quantities can be grouped in a logical order to constitute a system and that, further, systems of units have a separateness, a history and a fascination all their own if studied as such, instead of at random or in a haphazard fashion, being explained only as quantities met in the subject under study. This book sets out to study engineering quantities from a 'units approach' rather than from the standpoint of a particular subject.

Accordingly it is first necessary to define what is meant by a *system of units*.

1.2 A SYSTEM OF UNITS

The dictionary defines a 'unit' as 'any known determinate quantity by constant application of which any other similar quantity may be measured.' The word *determinate* has been derived from the verb *to determine*, one meaning of which is to *define*. Accordingly, we may regard a unit as a defined quantity. By using defined quantities other quantities of the same physical nature may be discussed and measured in terms of them, *e.g.* a mass which is, say, five times as big as a *one pound* mass is a *five pound* mass and so on. It is, of course, essential to define unit sizes so that engineers or scientists throughout the world can communicate their theories and discoveries concerning the nature of things without confusion.

To return to the dictionary: the word 'system' is defined as a 'regular order of connected parts'. An examination of units in

1

present day use shows that all units may be derived from a basic number of fundamental units and the derivation may be laid out in logical order.

So a system of units means 'a logical order of defined quantities by constant application of which other similar quantities may be measured'.

1.3   ABSOLUTE AND GRAVITATIONAL SYSTEMS: COHERENCE

Any unit which is by definition independent of changing physical quantities is called an absolute unit. Systems with a preponderance of such units are called absolute systems in the text.

Certain units based on the acceleration due to gravity (g, which varies slightly) are not absolute and are referred to as gravitational units. Systems with such units are generally referred to in this book as gravitational systems, although it should be pointed out that certain of the units within such a gravitational system may in fact be absolute themselves, *e.g.* the units of length and time in the British System are themselves absolute, but the system as a whole contains so many units derived from the gravitational unit, the pound force, that the system may on the whole be considered 'gravitational'.

The word 'coherent' as applied to units means that the constant connecting the unit with the quantities from which it is derived is unity. For example, in deriving the unit of power, the unit of work (energy) is divided by the unit of time giving, in the MKS Absolute System, one watt = one joule per second and, in the British Gravitational System, one horsepower = 550 foot-pound force per second; the watt is a coherent unit, the horsepower is not. Again systems with a preponderance of such units are called 'coherent systems' within the text, but it should be realised that the word 'coherent' used strictly applies to a particular unit rather than a system. This topic is considered fully in Chapter 3.

1.4   A DECIMAL SYSTEM OF UNITS: THE MKS SYSTEM

A Frenchman, Simon Stevin (1548–1620), is credited with the conception of the idea of a decimal system of units and the matter was given serious consideration by the French Académie des Sciences, which was founded in 1666, but it was not until after the French Revolution that formal proposals were submitted for a system destined to replace all other systems. The French scientists proposed a system based on three principles:

(1) that the basic standards should be taken from natural phenomena and not be based on arbitrary man-made standards;

(2) all units should be derived from a fixed number of fundamental units (three for a mechanical system, four for an electromechanical system);

(3) all multiples and subdivisions of units should be based on the decimal system (*see* Appendix I).

The three fundamental units were taken as those of *mass* (the *gramme*, then defined as the mass of a cubic centimetre of water at 4°C and 760 mm Hg), *length* (the *metre*, then defined as one ten-millionth part of the distance from pole to equator along the meridian passing through Paris) and *time* (the *second*, then defined as 1/86 400th part of the day and, later, of the mean solar day).

The *Metric* System, based on these quantities and incorporating the three basic principles, was decreed by law in France in 1795 and some 42 years later the use of all other systems was banned.

The Metre Convention, recognising the metric system, was signed by seventeen countries including the USA in 1875 and by the United Kingdom in 1884. Most countries, with the notable exceptions of the USA and the UK, proceeded eventually to recognise the system as the only legal one. The two exceptions went no further than making formal recognition of the system and allowing its use in commercial transactions.

The Bureau International des Poids et Mesures (BIPM) was established at Sèvres in France by the Comité International des Poids et Mesures (CIPM) and was to be responsible for the prototype reference standards (or *étalons*) and their safe keeping. [The CIPM now acts as a working committee for the Conference Générale des Poids et Mesures (CGPM) which meets every six years.]

Later the size of the standard mass measure was changed from the gramme to the *kilogramme* and, although the basic definitions of the three fundamental units have changed over the years to more accurately measurable and unchanging standards, the three basic units then adopted were in fact three of the basic ones of the present Système International d'Unités or International System (SI).

1.5   THE MKS GRAVITATIONAL SYSTEM AND THE CGS SYSTEM

Although the original proposals were in fact the basis of the present International System, the MKS Absolute System was displaced for many years by two 'offshoots', the MKS Gravitational System, which used the *kilogramme-force*, *i.e.* the force of gravity acting on the kilogramme mass as a third fundamental unit, and the CGS Absolute System using the centimetre and gramme as two of the basic units.

The MKS Gravitational System suffers the serious defect shared by all gravitational systems in that its unit of force, the kilogramme force or *kilopond*, is not an absolute quantity, *i.e.* independent of physical surroundings, but in fact varies with the value of the acceleration due to gravity (g, which varies slightly throughout the world). When the implications of this were fully appreciated some attempt was made to fix an arbitrary standard by defining a so-called standard value of g, but such a system is far from satisfactory compared with the present International System. The distinction between absolute and gravitational systems is explained in greater detail in Chapter 3.

The CGS System was adopted as a result of proposals made by the British Association for the Advancement of Science and became firmly established in electrical science and in physics in general after formal adoption by the first International Electrical Congress in 1881.

The electrical units based on the CGS System belonged either to the CGS Electromagnetic System, in which the permeability of free space was taken as unity, or the CGS Electrostatic System, in which the permittivity of free space was taken as unity (*see* Chapter 8 and Appendix III). Neither of these systems corresponded with the 'practical' system of electrical units (including the volt, ampere, joule and watt) and conversion factors involving powers of ten had constantly to be employed.

Electrical engineers have gone some of the way in remedying the consequences of this earlier choice, which may reasonably be considered to have been a *faux pas*, by being the first group of engineers to formally accept the MKSA System, now expanded into the International System (SI).

It may, however, be a little time before the old CGS Systems, especially the electromagnetic units (gauss, oersteds, etc.) still fervently held on to by certain of the magnetic materials specialists, are completely displaced.

1.6 IMPERIAL SYSTEMS OF UNITS

Systems of units in common use in the United Kingdom, parts of the British Commonwealth, and to some extent in the United States of America, may be considered to be based on the *pound* as the unit of mass, the *foot* as the unit of length and the *second* as the unit of time;* the mass and length units being a heritage of the Roman occupation. The pound mass is currently defined in relation to the kilogramme, and the foot in the United Kingdom in relation to the Imperial Standard Yard and in the United States relative to the US inch, which is in turn related to the metre (*see* Chapter 2). There are other slight differences in the British and American systems including the unit of mass, the 'ton' (2 240 lb in the UK, 2 000 lb in the USA), and the liquid measure, the 'gallon' (approximately 277 cubic inches in the UK, 231 cubic inches in the USA).

As with metric-based units there is also the problem of 'absolute' and 'gravitational' systems, the most commonly used, especially by mechanical engineers, being the gravitational system, in which a confusion, which may almost be regarded as inherent, between mass and weight leads to misuse of unit names and misunderstanding of derived quantities.

A problem peculiar to Imperial units is their non-decimal ratio between multiples and sub-units and factors of 3, 12, 36 and even 1 760 and 550 are encountered.

All in all the Imperial Systems have little to recommend them other than the fact that they are already in use and it will be costly to convert. Nevertheless, within the next few years the present Imperial Systems will be superseded by the International System in engineering and in other commercial uses. The reasons will become apparent in more detail as the text is studied.

1.7 THE MKSA SYSTEM: THE INTERNATIONAL SYSTEM

In 1904, at a meeting of the International Electrical Congress, an Italian engineer Giovanni Georgi presented a paper showing that the 'practical' system of electrical units involving the volt and ampere, etc. tied in quite logically with a mechanical absolute system based

* It is argued in some texts that the Imperial Gravitational System to be consistent must be considered as having the force unit, the *pound force* (or *pound weight*), as a basic unit instead of the *pound mass* (and therefore the mass unit is *derived*), but the point, though no doubt valid, is considered a little academic for the scope of this text.

on the metre, kilogramme and second (as opposed to the centimetre, gramme, second).

It took some time for Georgi's proposals to be fully appreciated and accepted, but by 1927, when a sub-committee of the International Electrotechnical Commission was set up to examine the units of electrical engineering in common use, considerable interest was shown in the Georgi system based on the metre, kilogramme, second and one other of the electromagnetic units. The recommendations of Georgi were accepted in 1935 and in 1950 the fourth basic unit (the ampere) was formally adopted giving the MKSA System.

At its tenth meeting in 1954, the CGPM adopted a rationalised (*see* Appendix III) and coherent MKSA System based on the four units indicated, together with the degree Kelvin as the unit of temperature and the candela as the unit of luminous intensity (Chapters 5 and 6 respectively). The title 'Système International d'Unités' was adopted at the eleventh meeting in 1960 (abbreviated SI in all languages).

The above historical notes apply to a system of units for use in electrical engineering. For general engineering use the pace of the move towards international acceptance of the SI, especially the MKS-based units of mechanics, varies from country to country. In 1958 Germany made the first serious move towards replacing their popular system, the MKS Gravitational System, and other European countries are following suit. France went as far as declaring the use of *all* other systems but the SI to be illegal from January 1st, 1962.

The United Kingdom is now taking serious steps towards general adoption of the system and the British Standards Institution is playing a major part in facilitating the change.

Useful publications at present available in the United Kingdom include PD 5686 'The Use of SI Units' and BS 3763 'The International System (SI) of Units', both published by the British Standards Institution, London. Formal definitions of the fundamental quantities are given in these publications and in the text as each quantity is dealt with.

The reader might also like to note that there is available from the publishers a text presented in 'programmed learning' form entitled 'Fundamental Electrical Quantities in SI Units' by the same author.

CHAPTER TWO

Mass, Length, Time, Velocity and Acceleration

2.1 MASS, LENGTH AND TIME

As discussed briefly in Chapter 1, a system of units may be based on a small number of fundamental units; a system for mechanics requires three. The three chosen as *intuitively fundamental* are *mass*, *length* and *time*. As has been indicated, the most common combinations upon which systems are based are the *centimetre, gramme, second* (CGS), the *metre, kilogramme, second* (MKS) and the *foot, pound, second* (FPS). (A system *can* be based on any other three quantities; for example, it can be argued that the MKS Gravitational System has the unit of force, the kilopond, as a basic unit and because the FPS Gravitational System does *not* have the unit of force, the pound force, as one of its basic units it is not in fact *one system* but a mixture of two. However, the fact remains that the quantity 'force' is not *intuitively fundamental, i.e.* one cannot visualise it as easily as mass, length or time.)

As was stated in Chapter 1, the formal definition of the gramme was first taken as the mass of a cubic centimetre of water at a temperature of 4°C and a pressure of 760 mm Hg. The formal standard of mass was later taken as the mass of a cubic decimetre (*i.e.* 10^3 cm^3) at 4°C and normal atmospheric pressure, *i.e.* the kilogramme.

It can be seen that the mass unit, since it depends upon the mass of a given volume, requires an accurate and reproducible measure of length, *i.e.* it depends to a certain extent upon the length unit and its definition (but note that 'depends upon' does *not* mean 'is derived from' so that mass is *not* a quantity derived from length). The original definition of the MKS length unit, the metre, was one ten-millionth part of the distance from pole to equator along the meridian passing through Paris. This definition also covers the definition of the centimetre since the latter unit is *one-hundredth* of one metre.

Prototype standards of mass and length, called étalons, were carefully made up and deposited at the BIPM at Sèvres. Despite the

7

care taken it was later found by further measurements of even greater accuracy that the length étalon was 0·2 mm less than the defined metre and the mass étalon was 28 mg greater than the defined kilogramme. It was realised that further measurements of the distance from pole to equator might yet again change the size of the basic units and so the size of the étalons were taken as the new basic units and the formal definitions abandoned.

The kilogramme mass unit is still the same today in its new role as one of the basic quantities of the SI and is formally defined as the mass of the international prototype which is in the custody of the BIPM at Sèvres (Third CGPM, 1901). A multiple of the kilogramme, the metric *tonne* (equal to 1 000 kg), will very probably be in common use once the SI is firmly established in everyday use.

However, the metre unit definition has been improved and taken closer to the original premise of the French Academy of Science that fundamental units should be independent of man-made standards, and is now defined as 1 650 763·73 wavelengths *in vacuo* of the radiation corresponding to the transition between energy levels $2 p_{10}$ and $5 d_5$ of the krypton-86 atom (Eleventh CGPM, 1960). Put more simply, it is defined in terms of the wavelength (*see* Chapter 6 for an explanation of this term) of the energy radiated when an electron changes specified energy levels in a certain atom. This definition is, of course, totally independent of any man-made standard and is accurately reproducible throughout the world.

The basic time unit, the second, was originally defined as 1/86 400th part of the day, and, later, the mean solar day. It was then taken as '1/31 556 925·974 7 of the tropical year for 1900, January 0 at 12 h ephemeris time' (Eleventh CGPM, 1960). The tropical year is the time interval between consecutive passage, in the same direction, of the sun through the earth's equatorial plane. The second is now defined in terms of an atomic period of vibration as the duration of 9 192 631 770 periods of the radiation corresponding to the transition between the two hyperfine levels of the ground state of the caesium-133 atom.

The Imperial units of length and mass are the foot and pound respectively. The foot is defined as one-third of the length at 62°F between the centres of two gold plugs in the bronze prototype Imperial standard yard (kept by the Board of Trade in London). The subunit, the *inch* (being 1/12th part of the Imperial foot), is in fact two parts per million less than 25·4 mm. The US inch is defined as 1/39·7th of the metre; it is in fact two parts per million greater than 25·4 mm. However, for calibration purposes the equivalence between the inch and the metric unit is taken as 25·4 mm exactly after an agreement between appropriate authorities in the USA, the UK, the British Commonwealth and also South Africa. The multiple unit the *yard* (yd) is equal to 3 feet.

The pound is the mass of the Imperial standard pound, the prototype kept with that of length. It is interesting to note that the Weights and Measures Act of 1878 totally confuses the concepts of 'mass' and 'weight' resulting in a legal definition of the 'pound' as a unit of weight, whereas in fact what was meant is a unit of mass (mass and weight are dealt with in Chapter 3). The kilogramme is now taken as 2·204 62 pounds (often approximated to 2·2 if accuracy requirements allow). Multiple units of the pound mass are the ton, which is equal to 2 240 lb, and the hundredweight (abbreviated cwt), which equals 112 lb.

The Imperial unit of time is of course the internationally accepted unit, the second.

2.2 SPEED AND VELOCITY

Any body which changes its position is said to be in motion. In considering the path taken by a body in motion (unless it is spinning) it is usual to take the trajectory of the centre of gravity of the body. The speed of any body is the rate of change with respect to time of the distance along its path from a fixed point on the path. The path may be straight or curved and the distance is the length *along the path* and not necessarily the shortest distance between the point at which the speed is measured and the fixed or *datum* point.

The speed of a body may be constant (*i.e.* the distance travelled per unit time is the same for all time intervals during the motion) or variable.

In stating the speed at a particular point, either in time or along its path, of a body having variable speed we are stating the distance which would be travelled in unit time immediately following the point at which the speed is being considered provided the speed stayed at that particular value.

Mathematically, the rate of change of distance s with time t is written ds/dt (ds by dt) and this symbol is used as a general one for speed whether it is variable or constant. If v is used as a symbol for speed then

$$v = \frac{ds}{dt} \tag{2.1}$$

The symbol v is the initial letter of the term 'velocity'. *Velocity* is the speed of a body in a *given direction, i.e.* it is a *vector* quantity and may be illustrated graphically by a vector diagram. It is not correct to use the word 'velocity' without specifying the *direction* in which the body is travelling. Numerically, of course, the velocity and speed of any body have the same value and the two quantities are measured in the same units.

The unit of speed and of velocity is the unit of length (or distance) per unit time, *i.e.* ft/s, m/s and so on. In many practical cases multiples of these units are used purely as a matter of convenience to give miles per hour (mile/h) and kilometres per hour (km/h). Knowing the connecting constants it is an easy matter to change units from one to another.

Equation (2.1) is the general expression for determining speed (or velocity) and the 'd' symbol implies the operation of the differential calculus, a branch of mathematics concerned in general with rates of change of a variable quantity with another. It is not correct to say, using the symbols given above, that

$$v = \frac{s}{t} \tag{2.2}$$

unless it is known that the speed of the body under consideration is constant.

The *average* speed of a body over a particular distance and in a given time is the value of the speed which would have to be maintained at a constant level for the body to cover the given distance in the given time.

Example 2.1

Calculate the average speed in mile/h and in ft/s of a car which covers a distance of 158 400 ft in 1 h.

In 1 h the car travels 158 400 ft, *i.e.* in 60 × 60 s the car travels 158 400 ft. Therefore

$$\text{the average speed} = \frac{158\ 400}{3\ 600}\ \text{ft/s}$$

$$= 44\ \text{ft/s}$$

$$= 158\ 400\ \text{ft/h}$$

$$= \frac{158\ 400}{1\ 760 \times 3}\ \text{mile/h}$$

(there are 1 760 yd to the mile and 3 ft to 1 yd)

$$= 30\ \text{mile/h}$$

Example 2.2
Calculate the distance in feet travelled by a train moving at a constant speed of 60 mile/h in 10 s.

$$60 \text{ mile/h} = \frac{60}{3\ 600} \text{ mile/s}$$

$$= \frac{60 \times 1\ 760 \times 3}{3\ 600} \text{ ft/s}$$

$$= 88 \text{ ft/s}$$

Therefore, distance travelled in 10 s = 880 ft.

Example 2.3
(For readers with a knowledge of differential calculus.)
The distance s metres travelled by a certain body from a fixed point is connected to the time t in seconds (measured after the beginning of the motion) by the equation

$$s = 3t^2 + 2t$$

  (i) Calculate the speed (a) after 3 s, (b) after 10 s.
  (ii) Calculate the distance in metres after 3 s and hence the average speed over this distance.

Using the calculus and differentiating the equation we get

$$\frac{ds}{dt} = 6t + 2$$

i.e.

$$v = 6t + 2$$

(i)(a) When

$$t = 3 \text{ s}$$

$$v = 20 \text{ m/s}$$

  (b) When

$$t = 10 \text{ s}$$

$$v = 62 \text{ m/s}$$

(ii) After 3 s the distance covered (using the equation connecting distances with time t given above)

$$= (3 \times 9) + (2 \times 3)$$

$$= 27 + 6$$

$$= 33 \text{ m}$$

Hence the average speed, given by distance/time, is

$$= \frac{33}{3} \text{ m/s}$$

$$= 11 \text{ m/s}$$

This example concerns a body travelling at a speed which is changing [shown by part (i)]. Note that the expression s/t gives an *average* speed or the constant speed at which the body would have had to travel to cover 33 m in 3 s. In fact, for part of this time the body travelled at less than 11 m/s and for the remainder at more than 11 m/s, *i.e.* it was accelerating (*see* the next section). The body was travelling at an instantaneous speed of 11 m/s at a time given by putting $v = 11$ into the equation relating v and t, *i.e.*

$$v = 6t + 2$$

$$11 = 6t + 2$$

$$t = 9/6$$

$$= 3/2 \text{ s}$$

but at all other times its speed was either less or more than this 11 m/s. Equation (2.1) should always be used unless other information indicates that the use of (2.2) would be acceptable.

Example 2.4
A car travels from A to B at an average speed of 30 mile/h, from B to C at an average speed of 40 mile/h and from C back to A at an average speed of 50 mile/h. The points A, B and C are equidistant. Calculate the average speed over the whole journey.

Let x be the distance in miles between A and B (and thus between B and C and between C and A). Then the time taken

$$\text{between AB} = \frac{x}{30} \text{ hours}$$

$$\text{between BC} = \frac{x}{40} \text{ hours}$$

$$\text{between CA} = \frac{x}{50} \text{ hours}$$

Hence the total time

$$= \frac{x}{30} + \frac{x}{40} + \frac{x}{50} \text{ hours}$$

$$= x \left(\frac{20 + 15 + 12}{600} \right) \text{ hours}$$

$$= \frac{47x}{600} \text{ hours}$$

Now total distance $= 3x$ miles. Therefore, the overall average speed

$$= \frac{\text{total distance}}{\text{total time}}$$

$$= \frac{3x}{47x/600} \text{ mile/hour}$$

$$= \frac{1\,800}{47} \text{ mile/hour}$$

$$= 38 \cdot 3 \text{ mile/hour}$$

Notice that in this example the overall average speed is *not* given by the average of the three average speeds, *i.e.*

$$\frac{30 + 40 + 50}{3} = 40 \text{ mile/hour}$$

This is because it is the distances covered in each case which are equal and not the respective times taken. The student is advised to rework the example assuming unequal distances but equal time intervals (from A to B, B to C, C to A).

2.3   ACCELERATION AND RETARDATION

When a body in motion is travelling at variable speeds the rate of change of speed with time at any point of the journey (either in time or along the path of motion) is called the *acceleration* if the speed is increasing and the *retardation* if the speed is decreasing.

The unit of acceleration (or retardation) is the unit of speed per unit time, *i.e.* basically feet per second per second (ft/s^2) or metres per second per second (m/s^2). Other variations are encountered such as, in motoring for example, mile/h s and so on.

The same remarks apply to acceleration as to speed concerning whether the quantity is constant or variable. Constant or uniform acceleration means that over successive, equal, short intervals of time the speed changes by the same amount. Variable acceleration implies that the speed change in successive, equal time intervals is different.

If a is the symbol for acceleration, then using the symbols of eqn. (2.1) the relationship between acceleration, speed and time is

$$a = \frac{dv}{dt} \tag{2.3}$$

which implies

$$a = \frac{d}{dt}(v) = \frac{d}{dt}\left(\frac{ds}{dt}\right) = \frac{ds^2}{dt^2}$$

i.e. the second derivative of s with respect to time.

Example 2.5
A car accelerates uniformly from zero to 30 mile/h in 10 s. What is the acceleration in FPS units?

30 mile/h is 44 ft/s. Hence the acceleration is 44 ft/s in 10 s or 4·4 ft/s in 1 s, *i.e.* acceleration is 4·4 ft/s^2.

Example 2.6
The acceleration of the car in Example 2.5 falls to a constant 2·2 ft/s^2 in the 5-s interval following the speed reaching 30 mile/h. What is the speed after 15 s from start?

Speed is $44 + 5 \times 2\cdot2$ ft/s, *i.e.* 55 ft/s or 37·5 mile/h.

Example 2.7
A train travels with a uniform acceleration of 1 ft/s^2 for 15 s and is then retarded at 2 ft/s^2 for 22 s after which it is at rest. Find the initial speed of the train and the greatest speed at which it travels during the motion.

Since we are not told the train starts from rest this cannot be assumed in finding the maximum speed. It is necessary to work 'backwards' through the journey, *i.e.* to consider the latter part of the journey first.

If the train takes 22 s to come to rest after reaching the maximum speed, and its retardation is 2 ft/s^2, then the maximum speed, *i.e.* the initial speed over the last part of the journey, is 22×2 ft/s, *i.e.* 44 ft/s.

Before reaching this speed it was accelerated at 1 ft/s^2 over a period of 15 s, so that the change in speed in this time interval was 15 × 1 ft/s, *i.e.* 15 ft/s.

Hence the initial speed at the start of the journey is (44 − 15) ft/s, *i.e.* 29 ft/s.

Example 2.8
The equation connecting distance s metres with time t seconds for the motion of a particular body is

$$s = 3t^3 + 6t^2 + 4t$$

Calculate the speed of the body after 1 s and after 2 s, the acceleration after 1 s and the average acceleration for the time interval between $t = 1$ and $t = 2$.

By differentiation of $s = 3t^3 + 6t^2 + 4t$, speed $v = 9t^2 + 12t + 4$.
After 1 s
$$v = 9 + 12 + 4$$
$$= 25 \text{ m/s}$$
After 2 s
$$v = 9 \times 4 + 12 \times 2 + 4$$
$$= 64 \text{ m/s}$$

By a second differentiation of $s = 3t^3 + 6t^2 + 4t$, *i.e.* by differentiating $v = 9t^2 + 12t + 4$, acceleration $a = 18t + 12$. After 1 second $a = 30$ m/s^2 and after 2 seconds $a = 48$ m/s^2.

$$\text{average acceleration} = \frac{\text{speed change between } t = 1 \text{ and } t = 2}{\text{time}}$$
$$= \frac{64 - 25}{1}$$
$$= 39 \text{ m/s}^2$$

Example 2.9
Express 5 km/h s in ft/s^2 (1 ft = 0·304 8 m).

$$1 \text{ ft} = 0·304\ 8 \text{ m}$$
Therefore
$$1 \text{ m} = \frac{1}{0·304\ 8} \text{ ft}$$
$$= 3·28 \text{ ft}$$

Therefore

$$1 \text{ km} = 3\ 280 \text{ ft}$$

$$5 \text{ km/h} = 5 \times 3\ 280 \text{ ft/h}$$

$$= 16\ 400 \text{ ft/h}$$

$$= \frac{16\ 400}{3\ 600} \text{ ft/s}$$

$$= 4 \cdot 56 \text{ ft/s}$$

Therefore

$$5 \text{ km/h s} = 4 \cdot 56 \text{ ft/s}^2$$

PROBLEMS ON CHAPTER TWO

(1) Calculate the average speed of a body which travels from a point A 100 ft from a fixed point D to a point B 144 ft from D in 10 s in (a) ft/s, (b) mile/h. Assume A, B and D lie along a straight line DAB.

(2) An electric train covers a journey between stations in three stages. It is first uniformly accelerated from rest at $4 \cdot 4$ ft/s^2 over a period of 10 s, it then maintains constant speed for 4 min and finally it is retarded with constant retardation to a halt. The journey takes 4 min 30 s. Calculate (a) the steady speed at the end of the first stage (in mile/h), (b) the distance covered during the second stage (in miles), (c) the retardation over the final stage (in ft/s^2).

(3) Two cyclists A and B, leave the same place at the same time and travel the same journey over the same distance. The ratio of the time taken by cyclist A to cyclist B is 2/3. What is the ratio of the average speeds of cyclist B to cyclist A? If the distance covered is 15 km and the sum of the average speeds is 35 km/h find the time taken for the journey in hours.

(4) The speed/time table of a particular journey is as follows:

Time (s)	0	5	10	15	20	25	30
Speed (m/s)	14	18	22	28	28	20	15

(a) What is the average acceleration, (i) in the first 10 s, (ii) between 15 and 20 s, (iii) in the last 5 s? (b) How long from the start of timing would it take for the speed to reach zero if the retardation of the last 5 s was constant? (c) Is the acceleration between 0 and 10 s uniform?

(5) Express the following uniform accelerations in ft/s^2. (a) 10 mile/h in 20 s, (b) 100 ft/min in 3 s, (c) $412 \cdot 5$ mile/h in 5 min.

CHAPTER THREE

Force

3.1 INTRODUCTION

So far five quantities have been discussed. These were the three basic quantities, upon the units of which the system is based, and two quantities whose units were derived from the basic ones. The next derived unit is the unit of force. Firstly, it is necessary to consider the nature of that which is called 'force' and upon what factors its magnitude or effect depends.

3.2 THE WORK OF NEWTON

Sir Isaac Newton (1642–1727) was the founder of classical mechanics and was responsible for the original postulate of a force of gravity. Though it is now known through the work of Albert Einstein that Newtonian physics is only a part of the physics of the universe as a whole, much of his work still forms the basis of the laws of everyday mechanics. His laws of motion give a definition of the term force and a means of defining a unit of force.

3.3 NEWTON'S LAWS OF MOTION

(1) Force is that which changes or tends to change a body's state of rest or of uniform motion in a straight line.
(2) The force acting on a body is proportional to the mass of the body and to its acceleration.

3.4 DISCUSSION

From the first law it can be seen that a force acting on any body will change the velocity of the body and thus must produce an acceleration and, further, that *unless* a force acts on a body it will stay in the same state, *i.e.* either at rest or moving along a straight line with constant velocity. It is emphasised that 'force' in the preceding sentence means 'resultant force'. There may be many forces acting upon a body; the resultant force is the one equivalent force which will completely replace all the forces acting. Since this is not a work on mechanics it will suffice to consider only two forces acting in the same straight line.

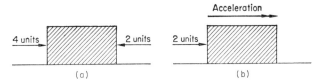

Fig. 3.1   Resultant force.

Figure 3.1(a) shows a body upon which two forces are acting in the same straight line and in opposite directions. Clearly, the force tending to move the body to the right, since it is greater, will have more effect than the force acting to the left. Nevertheless, the effect of the force acting to the right had the other force not been present is reduced by an amount equal in magnitude to the other force. Thus, as is shown in Fig. 3.1(b), the resultant force in this case will be one of $(4 - 2)$, *i.e.* 2 units, which will move the body to the right. The acceleration of the body due to the action of the resultant force (which is always in a direction along the line of action of the force) is, in this case, to the right.

From the second law of motion it can be seen that for a body of given mass the greater the force acting on it the greater will be the acceleration. Or, alternatively, for a body to be given a particular value of acceleration, the greater the mass of the body the greater will have to be the force to produce that acceleration.

3.5   THE EQUATION OF THE SECOND LAW

Suppose a force of F units acting on a body of mass M units gives it an acceleration of a units. The units of mass and acceleration are consistent with one another, *i.e.* are taken from only one system, *e.g.* mass M pounds and acceleration a feet per second per second for a system based on the foot, pound and second or mass M kilogrammes and acceleration a metres per second per second for a system based on the metre, kilogramme and second. From Newton's second law the quantities F, M and a are related by the statement of proportionality:

$$F \propto M \times a$$

which may be written as an equation by the insertion of a constant, thus:

$$F = k \times M \times a$$

where k is the constant.

3.6 THE VALUE OF THE CONSTANT: COHERENT SYSTEMS

Any value may be chosen for the constant; by choosing a particular number the unit of force is defined in terms of that number of mass–acceleration units. As was explained briefly in Chapter 1, it is often convenient to choose unity whenever possible for the value of constants in equations such as this. A unit defined by an equation in which the connecting constant is one is called a coherent unit. A system containing coherent units is called, naturally enough, a coherent system.

3.7 COHERENT UNITS OF FORCE

A coherent unit of force is defined as *one* mass–acceleration unit and may be derived from any system based on the three basic quantities already discussed, *i.e.* a system based on the foot, pound and second or the metre, kilogramme and second or the centimetre, gramme and second, and so on. Two such units will be discussed, that in a FPS-based system and that in a MKS-based system.

3.8 THE COHERENT UNIT OF FORCE IN THE FPS-BASED SYSTEM

The coherent unit of force in a system based on the foot, pound and second is called a *poundal*, abbreviated pdl. It should be noted that this unit is no longer in practical use. A force of *one poundal* exerted on a body of mass *one pound* will cause it to accelerate at *one foot* per *second* per *second*, *i.e.*

$$1 \text{ pdl} = 1 \text{ lb} \times \text{ft/s}^2$$

Example 3.1
What is the force acting on a body of mass 4 lb which is accelerating at 5 ft/s²?

$$\text{Force in poundals} = 4 \times 5 \text{ lb} \times \text{ft/s}^2$$

$$= 20 \text{ pdl}$$

Example 3.2
What will be the acceleration of a body of mass 3 lb when acted upon by a force of 10 pdl?

Since

$$\text{force (pdl)} = \text{mass (lb)} \times \text{acceleration (ft/s}^2)$$

$$10 = 3 \times \text{acceleration}$$

hence

$$\text{acceleration} = 10/3 \text{ ft/s}^2$$

3.9   THE COHERENT UNIT OF FORCE IN THE MKS-BASED SYSTEM (The International System)

The coherent unit of force in a system based on the metre, kilogramme and second is called a *newton*, abbreviation N. This is the unit of force in the International System of Units. A force of *one newton* exerted on a body of mass *one kilogramme* will cause it to accelerate at *one metre* per *second* per *second, i.e.*

$$1 \text{ N} = 1 \text{ kg} \times \text{m/s}^2$$

Example 3.3
A body of mass M kilogrammes is accelerating at 4·2 m/s² when the force exerted upon it is 14 N. What is the value of M?

Since

$$\text{force (N)} = \text{mass (kg)} \times \text{acceleration (m/s}^2)$$

$$14 = M \times 4\cdot2$$

hence

$$M = \frac{14}{4\cdot2} = 3\cdot33 \text{ kg}$$

3.10 ABSOLUTE SYSTEMS

One of the advantages of the coherent units of force so far developed is that they are independent of geographical location. Since the foot, pound and second or the metre, kilogramme and second are internationally defined without reference to any *changing* physical characteristics of the Earth, then the derived units of force are the same wherever the force is measured. Such a system containing units whose values are independent of physical quantities which are liable to change is called an *absolute* system.

Note that the word coherent implies only that the connecting constant between interdependent quantities is unity; it implies nothing as to the characteristic of the units, *i.e.* whether or not they are absolute, so that there may be any combination, *e.g.* coherent and absolute, incoherent and non-absolute and so on.

3.11 NON-ABSOLUTE SYSTEMS

In Sections 3.2–3.9 the equation of Newton's second law of motion was used to define a coherent unit of force. As was pointed out then it is possible to choose a value other than unity for the constant in the equation and still maintain the absolute nature of the unit; however, there is little point in doing so since unity is the most convenient value. Another type of system, the gravitational system, now to be considered, is a non-absolute system and also renders the value of the constant other than unity, *i.e.* it is incoherent as well as non-absolute. The unit of force in this system is based on the force of gravity.

3.12 THE FORCE OF GRAVITY

As was stated in Section 3.2, Sir Isaac Newton was responsible for the original postulate of the force of gravity. It is now known that *all* bodies exert a force of attraction upon each other. This force depends upon, amongst other things, the respective masses of the bodies concerned. For bodies of small mass the gravitational force between them may go unnoticed. However, the gravitational attraction between the Earth, which is of huge mass compared to any body upon it (about $13 \cdot 15 \times 10^{24}$ lb, in fact), and other bodies is very clearly felt; this special case of general gravitational force is called the force of gravity. The force of gravity acting upon any body depends upon its mass and the mass of the Earth. The force of gravity acting upon a body is also called the weight of the body. Since the weight of any body depends on the mass of the Earth as well as its own mass, it follows that if either of the masses changes then the weight also changes. The mass of the Earth does not change, of course, but, bearing in mind this fact concerning weight, it is easy to see that the weight of a particular body on Earth will not be the same as the weight of the same body on the Moon, for example, since the Moon has a mass approximately one-sixth that of the Earth. Strictly speaking, of course, now that travel to other parts of the universe is becoming a reality, weight should really be defined as the gravitational force between a body and whichever planet the body is situated on. For the moment, however, discussion will be confined to the force of gravity due only to the Earth.

Any unsupported body, *i.e.* without a restraining force, will move towards the Earth under the influence of its weight and will experience a definite acceleration.

3.13   THE ACCELERATION DUE TO GRAVITY, g

The acceleration due to gravity is the same for all bodies regardless of mass at the same place on the Earth. This was first shown by Galileo, who, in 1590, performed experiments involving the simultaneous release of two bodies of different mass from the top of the leaning tower of Pisa and observing the relative times at which they reached the ground. In fact they reached the ground simultaneously thus showing equal acceleration. This has, of course, been confirmed in a more refined manner since the time of Galileo.

The symbol for the acceleration due to the force of gravity is g. It will be noticed that the words 'at the same place on the earth' are included above. This is important since the value of g is only approximately the same at *different* places on the earth. The variation is only small but does nevertheless exist, and it is particularly noticeable between values of g measured over flat land and those measured over mountain ranges, and again between these values and those noted over deep oceans. The usual value of g is taken as $32 \cdot 18$ ft/s^2 or $9 \cdot 81$ m/s^2.

3.14   WEIGHT EXPRESSED IN COHERENT UNITS

The weight of a body is defined as the force of gravity which acts upon the body. Since weight is in fact a force, it can be expressed in force units of any system.

Example 3.4

Express the weight of a body of mass 1 lb in (a) poundals, (b) newtons (take g as 32 ft/s^2 or 9·8 m/s^2).

The weight of a 1 lb body is the force upon it due to gravity, *i.e.* the force which gives it an acceleration of g.

(a) Since the poundal is a coherent unit the following equation applies:

$$\text{force (pdl)} = \text{mass (lb)} \times \text{acceleration (ft/s}^2)$$

hence

$$\text{force} = 1 \times 32 \text{ pdl}$$

$$= 32 \text{ pdl}$$

(b) Since the newton is a coherent unit the following equation applies:

$$\text{force (N)} = \text{mass (kg)} \times \text{acceleration (m/s}^2)$$

and, since 1 lb mass is equal to 0·453 6 kg, then

$$\text{force} = 1 \times 0.453\ 6 \times 9.8 \text{ N}$$

$$= 4.445 \text{ N}$$

Thus, since weight is a force it can be expressed in coherent units. However, a new and *different* unit of force can be *defined* using weight, in particular the weight of unit mass. This unit is called the *gravitational* unit of force and may be defined in terms of the basic units of either the FPS-based system or the MKS-based system.

3.15 THE GRAVITATIONAL UNIT OF FORCE IN THE FPS-BASED SYSTEM

The gravitational unit of force in the FPS-based system is called the *pound weight* (abbreviated lb wt) or *pound force* (abbreviated lbf). The latter name is preferred. *One pound force* is the force of gravity experienced by a mass of *one pound* and since the force of gravity gives a body, whatever its mass, an acceleration of g, then *one pound force* gives a body of mass *one pound* an acceleration of g ft/s^2.

Recalling the FPS coherent unit of force (the poundal) and that *one poundal* gives a body of mass *one pound* an acceleration of 1 ft/s^2, it follows that 1 lbf is g times larger than 1 pdl since the acceleration it gives to the *same* mass is g times larger. This was in fact shown in Example 3.4(a) in which the value of g was taken as 32 ft/s^2.

3.16   THE GRAVITATIONAL UNIT OF FORCE IN THE MKS-BASED SYSTEM

The gravitational unit of force in the MKS-based system is called the *kilogramme force* (kgf) or *kilopond* (kp). One kilopond gives a body of mass one kilogramme an acceleration of g metres/(second)^2.

From above it follows that 1 kp is g times larger than the MKS coherent unit of force, the newton. It should be noted that g in this case is the m/s^2 value, *i.e.* 9·81, or as given and not, of course, the ft/s^2 value.

3.17   GRAVITATIONAL SYSTEMS OF UNITS: USE AND NATURE

Systems of units employing the gravitational unit of force, *i.e.* gravitational systems, contain units which depend on physical quantities which change slightly and hence are not absolute in the sense already defined.

Any force whether or not it is due to gravity may be measured in gravitational units and such a system may be used in exactly the same way as the coherent systems already discussed.

3.18   MEASURING ANY FORCE IN GRAVITATIONAL UNITS

To measure any force in coherent units it is only necessary to multiply together mass and acceleration (*i.e.* the constant in the equation of Newton's second law, Section 3.5, is unity). As has been shown, the gravitational unit of force for any system is g times larger than the corresponding coherent unit of force. It is therefore necessary to divide the product of mass and acceleration (which gives coherent units) by the correct value of g. In other words the constant in the equation of Newton's second law is no longer unity but is equal to 1/g. This is shown in another way below.

3.19   THE VALUE OF THE CONSTANT IN THE GENERAL EQUATION OF NEWTON'S SECOND LAW FOR GRAVITATIONAL UNITS

The general equation from Section 3.5 is

$$\text{force} = \text{k} \times (\text{mass}) \times (\text{acceleration})$$

and since one gravitational unit of force gives one unit of mass an acceleration of g units, then

$$1 = k \times 1 \times g$$

hence

$$k = \frac{1}{g}$$

So for force expressed in gravitational units the equation is

$$\text{force} = \frac{1}{g} \times (\text{mass}) \times (\text{acceleration})$$

Example 3.5

What is the force which gives a body of mass 24 lb an acceleration of 4 ft/s^2 expressed in (a) poundals and (b) pounds force?

(a) The poundal is a coherent unit; therefore the following equation applies:

$$\text{force (pdl)} = \text{mass (lb)} \times \text{acceleration (ft/s}^2)$$

hence

$$\text{force} = 24 \times 4 \text{ pdl}$$

$$= 96 \text{ pdl}$$

(b) The pound force is a gravitational unit; therefore the following equation applies:

$$\text{force (lbf)} = \frac{1}{g} \times \text{mass (lb)} \times \text{acceleration (ft/s}^2)$$

hence

$$\text{force} = \frac{24 \times 4}{32} \text{ lbf}$$

$$= 3 \text{ lbf}$$

It will be seen that a force of 3 lbf achieves the same effect as a force of 96 pdl (since they both give the same mass the same acceleration) so that 1 lbf is equivalent to 32 pdl (if g is taken as 32 ft/s^2).

Example 3.6

What is the force which gives a body of mass 10 lb an acceleration of 1 ft/s^2 expressed in (a) newtons and (b) kiloponds? (Take g as 9·8 m/s^2; 1 lb equals 0·453 6 kg; 1 ft equals 0·304 8 m.)

Notice that the mass here is given in pounds and the acceleration in ft/s². Two methods of solution are available. For the first method the force may be determined in pounds force and poundals and these figures may then be converted to newtons and kiloponds, but this method requires the ft/s² value of g which is not given directly (though it may be found by converting 9·81 m/s²) and also requires the relationship between English and metric force units which is also not given (though again it may be found—unit conversion is dealt with in greater detail in a later chapter). The second method involves changing both mass and acceleration to metric units before finding the force. This method will be used as it is the more direct (since conversion ratios are given).

(a) The mass of the body is 10 lb, *i.e.* (10 × 0·453 6) kg. The acceleration is 1 ft/s², *i.e.* (1 × 0·304 8) m/s². Since the newton is a coherent unit,

$$\text{force (N)} = \text{mass (kg)} \times \text{acceleration (m/s}^2)$$

hence

$$\text{force} = (10 \times 0\text{·}453\ 6) \times (1 \times 0\text{·}304\ 8)\ \text{N}$$

$$= 1\text{·}382\ \text{N}$$

(b) Since the kilopond is a gravitational unit,

$$\text{force (kp)} = \frac{1}{g} \times \text{mass (kg)} \times \text{acceleration (m/s}^2)$$

hence

$$\text{force} = \frac{1}{9\text{·}8} \times (10 \times 0\text{·}453\ 6) \times (1 \times 0\text{·}304\ 8)\ \text{kp}$$

$$= 0\text{·}141\ \text{kp}$$

Again it can be seen that the gravitational unit is 1/g times the coherent unit.

Example 3.7
Express the weight of a man of mass 160 lb in (a) poundals and (b) pounds force (take g as 32 ft/s²).

The weight of a body is the force it experiences due to gravity, *i.e.* that force which causes an acceleration of g, in this case 32 ft/s².

(a) The equation connecting force, mass and acceleration for the coherent unit of force, the poundal, is

$$\text{force} = (\text{mass}) \times (\text{acceleration})$$

The mass is 160 lb and the acceleration is 32 ft/s^2, hence:

$$\text{force} = 160 \times 32 \text{ pdl}$$
$$= 5\,120 \text{ pdl}$$

(b) The equation connecting mass, force and acceleration for the gravitational unit of force, the pound force, is

$$\text{force} = \frac{(\text{mass}) \times (\text{acceleration})}{g}$$

hence

$$\text{force} = \frac{160 \times 32}{32} \text{ lbf}$$

$$= 160 \text{ lbf}$$

It will be noticed that owing to the fact that g is in the numerator as well as in the denominator of this equation the weight of the man in gravitational units is *numerically* the same as the mass. An alternative way of observing that this must be so is as follows:

Weight is a force—the force due to gravity. Force is proportional to (mass) × (acceleration). The force of gravity produces the *same* acceleration for all bodies regardless of mass. Hence weight is proportional to mass (true whichever units are used). Using gravitational units the weight of a body of mass *one* pound is *one* pound force. Thus the weight of a body of any mass *M* pounds is *M* pounds force. Thus the weight of a body expressed in gravitational units has the same *numerical* value as the mass of the body.

This numerical equality does not occur when coherent units are used since the weight of a body of unit mass is not one unit of force but g coherent units of force (hence the weight of a body of any mass *M* units will be *Mg* coherent units of force).

3.20 THE CONFUSION BETWEEN MASS AND WEIGHT

The numerical equality between mass and weight when weight is expressed in gravitational units leads to a confusion between these quantities. This confusion is increased when using the Imperial Gravitational System by the fact that the name of the unit of mass and the name of the unit of force both contain the word 'pound'. (In the Metric Gravitational System the unit of force can be called the kilogramme force or kilogramme weight, but the name kilopond is generally used.) Often when using the Imperial Gravitational

System the additional word 'force' is left off the name of the force unit and it is by no means unusual to hear 'a weight of x lb . . .' or 'a body weighs x lb . . .' This is, of course, quite wrong. Weight is force and must be expressed in force units. If it is necessary to use the word 'weight' as a noun, *i.e.* describing a mass of standard and accurate value for use with balances, etc., one should endeavour to use the word 'mass' in conjunction with it (a 'weight' of mass 1 lb, etc.), or better still to avoid the use of the word 'weight' altogether in this connection and retain it only when the force due to gravity is being described. The statement 'a body weighs x lb . . .' is the most incorrect use of all; it should read 'a body weighs x lbf . . .'. The word 'force' must be kept in the name of the Imperial gravitational unit of force in order to avoid confusion, not only when force is being discussed but also when other quantities whose units are derived from, or contain, the force unit are under consideration. These quantities will be discussed in succeeding chapters.

3.21   g—TO MULTIPLY OR DIVIDE?

A problem particularly encountered by students when using the standard equations for derived quantities is whether to multiply or divide by g—or indeed if g should be included at all. The fundamental unit of force—which effectively determines the type of system—should first be recalled and the fact that one gravitational unit of force is equal to $1/g$ of one coherent unit of force should be borne in mind. Understanding the basic essentials of a unit system avoids confusion. As other derived units are developed the question of 'what to do with g' will be considered as it arises.

3.22   REASONS FOR THE ADOPTION OF THE SI

As has been stated, the International System of Units (SI) is being adopted for technical and commercial use in the United Kingdom and throughout Europe. The timetable varies for each section of industry but it is anticipated that by 1975 the Imperial System, based on the pound, foot and second, will no longer be in use in the UK and the gravitational MKS System (using the kilopond) will be completely displaced on the Continent.

The reasons for the changeover may be summarised as follows:

(1) The International System is an absolute system, *i.e.* its basic units are independent of the location of their measurement and do not depend on variable gravitational forces.

(2) The International System is mainly coherent, *i.e.* units derived from others are obtained by direct multiplication or division and connecting constants are unity.

(3) There is no confusion between the SI units of force (the newton) and mass (the kilogramme).

(4) Multiples and sub-multiples of SI units are of decimal form, *i.e.* 10, 10^2, 10^3, etc. or 10^{-1}, 10^{-2}, 10^{-3}, etc. (*see* Appendix I).

(5) The expansion of the SI Mechanical System with the additional basic unit, the coulomb, leads to the 'practical' electrical system using units long accepted as being of reasonable size, *i.e.* volt, ampere, watt, etc.

It can be seen from the above that the SI is a natural choice for many reasons, including ease of measurement, dependability of standards, practicality of unit sizes and so on. It will eventually supersede all other systems if present trends continue.

3.23 SUMMARY OF FORCE UNITS

The following is a summary of the points made in this chapter concerning the quantity force:

(1) The third derived unit is the unit of force. Since the remaining units in a mechanical system of units include the force unit, the choice of the unit of force will determine to some extent the type of system.

(2) An absolute system of units contains quantities which depend on *unchanging* physical quantities.

(3) Force is defined as that which changes or tends to change a body's state of rest or of uniform motion in a straight line.

(4) Force is related to the mass and acceleration of the body upon which it is acting by the general equation of Newton's second law: force = k × (mass) × (acceleration), where k is a constant. The choice of unit determines the value of k. If k is unity the unit is *coherent*.

(5) Absolute coherent systems discussed include:

(i) The FPS-based system in which the unit of force, the *poundal*, when acting on a body of mass *one pound*, gives it an acceleration of ONE *foot* per *second* per *second*. This system is not in practical use.

(ii) The MKS-based system in which the unit of force, the *newton*, when acting on a body of mass *one kilogramme*, gives it an acceleration of *one metre* per *second* per *second*. This system is the basis of the International System now being adopted in the UK and throughout Europe.

(6) A *gravitational* force of attraction exists between all bodies. The gravitational force existing between the Earth and any body, which is a special case of the general case of gravitation, is called *the force of gravity*. All bodies regardless of mass when free to move under the influence of the force of gravity will move towards the Earth with an acceleration denoted by g. The acceleration due to gravity, g, is the same for all bodies, whatever their mass, *at the same place upon the Earth*. The value of g varies slightly from one place to another on the earth.

(7) The *weight* of a body is the force of gravity acting upon it. Since weight is a force it may be expressed in the force units of any system.

(8) In a gravitational system of units the force of gravity acting on a body of mass 1 unit, *i.e.* the *weight* of a body of mass 1 unit, is taken as the unit of force. Thus, in the FPS-based system the unit of force is the *pound force* and in the MKS-based system it is the *kilogramme force* or *kilopond*.

(9) It follows that the constant in the general equation of Newton's second law, (force) = (constant) × (mass) × (acceleration), is 1/g. This means that the unit of force in a gravitational system is *not* coherent.

(10) Since the gravitational unit of force depends upon physical quantities which are *not* unchanging then the unit is also *non-*absolute.

(11) Gravitational systems discussed were:

(i) The FPS-based system in which the unit of force, the *pound force*, when acting on a body of mass *one pound*, gives it an acceleration of g *feet* per *second* per *second*. (g is of the order of 32 and is constant at a given place upon the Earth. Its value for a particular problem is usually given.) This system is at present in use in Canada, the USA and certain other countries. It is being replaced in the UK by the International System.

(ii) The MKS-based system in which the unit of force, the *kilopond*, when acting on a body of mass *one kilogramme*, gives it an acceleration of g *metres* per *second* per *second*. (g is of the order of 9·8 and the same conditions as above apply.) This system was widely used in Europe until recently. It is being replaced by the International System.

(12) The most useful system of units is undoubtedly the International System based on the metre, kilogramme and second. This system

is coherent, absolute and leads naturally to practical units, some of which have been in use for many years.

3.24 PRESSURE

Pressure is defined as force per unit area. The units are fairly straightforward, being the pound force per square foot (lbf/ft^2) or per square inch (lbf/in^2) for the Imperial Gravitational System, and the kilopond per square metre (kp/m^2) or square centimetre (kp/cm^2) for the Metric Gravitational System.

The SI unit is, of course, the newton per square metre (N/m^2), the most probable name to be adopted for this unit being the *pascal*. At the time of writing this has yet to be decided.

3.25 WORKED EXAMPLES ON CHAPTER 3

(1) Determine the tension in a rope suspending a body of mass 10 kg. Express the force in kiloponds and newtons (g = 9·8 m/s²).

Solution
There are two forces acting on the body—the weight of the body acting downwards and the tension in the rope acting upwards. Since the body is not moving there is no *resultant* force. Thus, the tension in the rope and the weight of the body must be equal and opposite:

$$\text{weight of the body} = 10 \text{ kp}$$
$$= 10 \times 9\cdot8 \text{ N}$$
$$= 98 \text{ N}$$

thus the tension in the rope = 98 N or 10 kp

(2) Two bodies of masses 50 and 40 lb respectively are connected by a rope which is hung over a smooth pulley. Determine the tension in the rope (in lbf) and the magnitude and direction of the resultant acceleration (take g as 32 ft/s²).

Solution
The bodies are situated as in Fig. 3.2. Let the tension in the rope be T pounds force acting in the directions shown and let the resultant acceleration be a feet/(second)² in the direction shown (if this direction is incorrect the answer for the acceleration will be negative). Since

the bodies are moving there *is* a resultant force on each body and the rope tension will be neither 40 or 50 lbf. The tension is, of course, constant throughout the rope since the pulley is smooth and there are no forces due to friction.

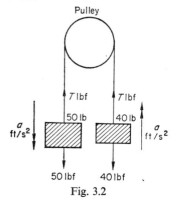

Fig. 3.2

For the 50 lb mass the resultant force *in the direction of the acceleration* is $(50 - T)$ lbf. Hence, for the 50 lb mass, using the equation of Newton's second law for gravitational units,

$$(50 - T) = \frac{50 \times a}{32} \tag{3.1}$$

Similarly, for the 40 lb mass,

$$(T - 40) = \frac{40 \times a}{32} \tag{3.2}$$

There are two unknowns and two equations.
From eqn. (3.1):

$$\frac{50 - T}{50} = \frac{a}{32}$$

From eqn. (3.2):

$$\frac{T - 40}{40} = \frac{a}{32}$$

Therefore

$$\frac{50 - T}{50} = \frac{T - 40}{40}$$

$$2\,000 - 40T = 50T - 2\,000$$

$$4\,000 = 90T$$

$$T = 44 \cdot 44 \text{ lbf}$$

(notice that T is between the values 40 and 50).

Also from eqn. (3.1)

$$a = \frac{32}{50} (50 - 44\cdot44)$$

$$= \frac{32}{50} \times 5\cdot56$$

$$= 0\cdot358\ 4\ \text{ft/s}^2$$

and the direction assumed was correct since the sign is positive.

(3) A lift carrying bodies of total mass 1 600 lb is accelerated upwards at 2 ft/s². Find the total upward thrust in pounds force (take g as 32 ft/s²).

Solution
The bodies of mass 1 600 lb experience a downward force due to gravity, *i.e.* their weight, of 1 600 lbf. If no other forces act the bodies would fall towards the Earth. As was seen in the worked Example (1) an upward force equal to their weight would maintain the bodies at rest. However the lift in this problem is accelerating upwards at 2 ft/s². Hence, the total upward thrust must be composed of a force sufficient to prevent falling, *i.e.* equal to the weight of the lift, and a force sufficient to give this mass an upward acceleration of 2 ft/s².

Total thrust = (force to overcome that due to gravity)
                              + (force to give acceleration)

$$= 1\ 600 + \frac{1\ 600 \times 2}{32}\ \text{lbf}$$

$$= 1\ 700\ \text{lbf}$$

(4) A train of mass 150 tonnes is accelerated at 2 m/s² by a total tractive force of 320 000 N. Find the resistive force.

Solution
The tractive force to give a mass of 150 tonnes, *i.e.* 150 × 1 000 kg, an acceleration of 2 m/s² is given by:

$$\text{tractive force} = 150 \times 10^3 \times 2\ \text{N}$$

$$= 300\ 000\ \text{N}$$

Now, the *total* tractive force = 320 000 N

and since

total tractive force = (force to overcome resistance)

+ (force to give acceleration)

then

$$\text{resistive force} = 320\,000 - 300\,000 \text{ N}$$
$$= 20\,000 \text{ N}$$

PROBLEMS ON CHAPTER THREE

(Take g as 32 ft/s² or 9·8 m/s².)

(1) Find the acceleration when a force of (a) 2 lbf acts on a 5 lb mass, (b) 10 pdl acts on a 6 lb mass, (c) 10 kp acts on a 2 kg mass, (d) 24 N acts on a 2 kg mass.

(2) Find the force acting in the following cases (in the units shown): (a) 3 lb mass accelerating at 3 ft/s² (in lbf and in pdl), (b) 5 kg mass accelerating at 2 m/s² (in N and in kp).

(3) A body of mass 10 lb falls under gravity. What will be the acceleration assuming the resistive forces to be constant at 10% of the downward force?

(4) The gravitational attraction between the Moon and any body may be taken as one-sixth that between the Earth and the same body. (a) What is the weight of a body of mass 24 lb on the Moon? (b) What will be its mass on the Moon? (c) What will be the equation connecting force (lbf), mass (lb) and acceleration (ft/s²) when applied to problems in mechanics on the Moon? (Take g on Earth as 32 ft/s².) (d) Why is it not advisable to use gravitational systems of units when solving problems in mechanics concerned with space research?

(5) A 300 kg van accelerates uniformly from rest to a speed of 45 km/h in 10 s. Calculate the resultant force (in newtons) acting on the van.

(6) A fully laden lift of 1 000 kg capacity is accelerated at 1 m/s². Find (a) the resultant upward thrust (in kN) and (b) the total upward force acting on the lift (in kN).

(7) An electric train makes the run between two stations in three stages. The first stage takes 2 min 12 s, during which time the train accelerates uniformly from rest to a steady speed. The second stage lasts 5 min during which time the train covers 3¼ miles at constant speed. During the third stage the motors are cut and the train is

allowed to come to rest under the action of the resistive force which may be assumed constant at 4 480 lbf. The mass of the train is 192 tons. Determine the tractive effort made by the motors during the first stage and the time taken for the train to come to rest from the instant the motors are disconnected.

CHAPTER FOUR

Energy, Work, Power and Torque

4.1 INTRODUCTION

So far three quantities derived from the fundamental ones have been discussed—velocity, acceleration and force. The next three derived quantities to follow logically from the concept of force are energy, work and power. As usual, precise and accurate definitions must be formulated if ambiguity and confusion are to be avoided. This is particularly so where quantities whose names are in everyday use are encountered.

The terms 'energy', 'work' and 'power' are frequently used in everyday language in instances where anything but the precise meaning of the word is in fact being described.

The three terms will be fully discussed and defined and their units stated with special emphasis on the units of the International System.

4.2 ENERGY

A comprehensive and accurate definition of energy may be stated as follows:

Energy is the ability to set up a force and in so doing to change, convert or modify a body's physical state, shape or mass or state of rest or of uniform motion in a straight line.

It will be seen that this definition embodies a definition of force (Chapter 3) and that, in fact, the force definition has been extended from the one already given to include both external and internal forces. The force set up whenever energy is expended (or, to be precise, converted) is external if it is directly observable inasmuch as the body upon which the force acts is seen to change its state of motion (or of rest); the force is internal if it is confined to the intermolecular or atomic structure of a material so that its presence is detected by observed physical changes of the material.

36

In order to set up a unit of energy it is necessary to define it by reference to accurately reproducible conditions. These conditions are better met by consideration of the external force set up by energy rather than an internal force, which though definitely in existence is not so easily observed or its effects measured. A measure of available energy in the setting up of such an external force is obtained by the magnitude of the force and the distance moved by the body upon which it acts. Consequently the product of the force size and the distance moved by the point of application of the force is taken as the unit of the energy expended (converted) in the setting up and movement of the force.

Thus we say that if a force of *one pound force* moves its point of application *one foot* the energy converted is *one foot pound force* and for this system of units (Imperial Gravitational) the energy unit is the *foot pound force*, abbreviated ft lbf.

The Metric Gravitational System (having the kilopond as the force unit and the metre as the length unit) has the energy unit *metre kilopond* (kp m).

For the Imperial Absolute Coherent System, in which the unit of length is the foot and the unit of force is the poundal, the energy unit is the *foot poundal* (ft pdl) and for the International System (MKS Absolute) the energy unit is the *metre newton* (N m). One metre newton is in fact given the special name *joule* after the scientist (*see* Chapter 5). The unit joule is abbreviated J.

Example 4.1

Calculate the energy converted if a force of 32 lbf moves a distance of 1 ft (a) in the units of the FPS Absolute System (take g = 32), (b) in the units of the FPS Gravitational System and (c) in the units of the International System.

(a) If g = 32 then 32 pdl = 1 lbf so that the force is 32 × 32 pdl, *i.e.* 1 024 pdl. The distance moved is 1 ft, hence

$$\text{(energy converted)} = \text{(force)} \times \text{(distance moved)}$$
$$= 1\ 024 \times 1$$
$$= 1\ 024 \text{ ft pdl}$$

(b)

$$\text{(Energy converted)} = \text{(force)} \times \text{(distance moved)}$$
$$= 32 \times 1$$
$$= 32 \text{ ft lbf}$$

From (a) and (b) and by simple consideration it is clear that if g is 32 then 1 ft lbf = 32 ft pdl, *i.e.* in general, as 1 lbf = g pdl then 1 ft lbf = g ft pdl.

(c) For the International System (MKS Absolute) the force unit is the newton and the length unit is the metre. It is therefore necessary to first determine the force (32 lbf) in newtons and the distance moved (1 ft) in metres in order to determine the energy converted in the units of this system. A method of conversion when only the ratios between the units of the three fundamental quantities mass, length and time in the two systems are known is given in Chapter 10. However, for the moment we will accept that 1 lbf = 4·45 N and 1 ft = 0·30 m, so that

$$
\begin{aligned}
(\text{energy}) &= (\text{force}) \times (\text{distance}) \\
&= 32 \ (\text{lbf}) \times 10 \ (\text{ft}) \ \text{ft lbf} \\
&= 32 \times 4{\cdot}45 \ (\text{N}) \times 10 \times 0{\cdot}3 \ (\text{m}) \ \text{N m} \\
&= 427{\cdot}5 \ \text{N m} \\
&= 427{\cdot}5 \ \text{J}
\end{aligned}
$$

The unit of energy based, as it is, on an external force moving its point of application may nevertheless be used to measure all kinds of energy even though the force is not directly observable as in the case of electrical energy (Chapter 7), heat energy (Chapter 5), light energy (Chapter 6) and so on.

In such cases in using the units derived from the movement of an external force we are in fact comparing the particular energy with that more directly observable.

4.3   TYPES OF MECHANICAL ENERGY

Energy loosely described as 'mechanical' may be divided into two broad classes, that of energy of movement or *kinetic energy* and that of energy by virtue of position or *potential energy*.

Of the two, kinetic energy is the easiest to visualise. It is clear that a moving body has energy and its capacity to set up a force is in no doubt when one considers the damage caused by traffic accidents! Potential energy exists whenever a force is already in existence and a restraining force is necessary to avoid the action of the already existing force on the body having the potential energy. One example of this is illustrated by considering an electron or other charged particle (*see* Chapter 7) situated near an electrically charged body. The force of attraction or repulsion already exists and the electron if not restrained will move under the influence of the force.

Another example, more commonly observed, is the action of the force of gravity (Chapter 3). All bodies near the Earth will move towards it under the influence of the force of gravity unless they are restrained, *i.e.* supported. The method of determination of the magnitude of kinetic energy and potential energy is dealt with in the following sections.

4.4 CALCULATION OF KINETIC ENERGY

The kinetic energy of any body of mass m and moving with velocity v may be determined using the equation

$$\text{kinetic energy} = \tfrac{1}{2} mv^2 \tag{4.1}$$

for coherent units (ft pdl, N m, etc.), or the equation

$$\text{kinetic energy} = \tfrac{1}{2} \frac{mv^2}{g} \tag{4.2}$$

for gravitational units (ft lbf, kp m, etc.).

These equations may be derived for the special case of uniform acceleration (Section 4.4A) or for the absolutely general case by the use of the intergral calculus (Section 4.4B).

4.4A Uniformly accelerating bodies

Consider a body of mass m accelerating uniformly over a period of time t from zero velocity to a velocity v (units are consistent, *i.e.* from one system).

The acceleration is v/t and the force causing movement is mv/t (for coherent systems). The distance moved during time t is $(v/2)t$, where $v/2$ is taken as the average velocity over period t.

Then the energy of the body at the end of the period (at which time the body is moving with velocity v) is given by (force) × (distance), *i.e.*

$$\frac{mv}{t} \times \frac{vt}{2} \text{ coherent units}$$

$$\textit{i.e. } \tfrac{1}{2} mv^2$$

4.4B The general case

The acceleration at any time is dv/dt (all symbols represent quantities as indicated in Section 4.4A). Hence force is

$$m \times \frac{dv}{dt} \text{ coherent units}$$

The distance moved in a very small interval of time dt is v(dt). Hence energy acquired during interval of time dt

$$= \text{(force)} \times \text{(distance)}$$

$$= \text{m} \frac{\text{d}v}{\text{d}t} \times v(\text{d}t)$$

$$= mv \, \text{d}v \text{ coherent units}$$

Therefore total energy over period t units of time (in which the velocity is increased from zero to v)

$$= \int_0^v mv \, \text{d}v$$

$$= \tfrac{1}{2} mv^2$$

To express absolute coherent units of force (poundals, newtons, etc.) in non-absolute gravitational units of force (pounds force, kiloponds, etc.), it is necessary to divide by g. Hence the expression $\tfrac{1}{2} mv^2$ for coherent absolute energy units must be divided by g to give gravitational units.

A check on the dimensions of $\tfrac{1}{2} mv^2$ and on energy units is discussed in Chapter 10.

Example 4·2
Calculate the kinetic energy in (i) coherent absolute units, (ii) gravitational units of the bodies in the following situations: (a) a body of mass 10 lb moving at 40 ft/s; (b) a body of mass 5 kg moving at 10 m/s.

(a) Mass 10 lb, velocity 40 ft/s.

(i)   kinetic energy $= \tfrac{1}{2} mv^2$

$$= \tfrac{1}{2} \times 10 \times 40^2$$

$$= 8 \, 000 \text{ ft pdl}$$

The unit is foot poundal since the coherent equation was used.

(ii)   kinetic energy $= \dfrac{8 \, 000}{32}$ ft lbf (taking g $= 32$)

$$= 250 \text{ ft lbf}$$

(b) Mass 5 kg, velocity 10 m/s.

(i)   kinetic energy $= \tfrac{1}{2} \times 5 \times 10^2$

$$= 250 \text{ N m or J}$$

The metre newton (or joule) is the coherent absolute unit.

(ii)   kinetic energy $= \dfrac{250}{9 \cdot 81}$ kp m (taking g $= 9 \cdot 81$)

                            $= 25 \cdot 5$ kp m

4.5 CALCULATION OF POTENTIAL ENERGY

Potential energy is the energy of a body by virtue of its position relative to a given level. To calculate the potential energy of any body it is usually convenient to find the energy used (given to the body) to put the body in that position from the given level. For example, the potential energy relative to the ground of a body of mass m, weight W, held h units of length above ground is calculated by determining the energy expended in raising the body to that height, in this case by multiplying the force necessary to lift it (equal and opposite to the weight W) by the distance (h), *i.e.* Wh units. Whether the units are coherent absolute or gravitational depends upon what force units the height is expressed in.

Example 4.3
Calculate the potential energy of a body of mass 20 lb held 10 ft above ground. Express the energy in absolute and gravitational units.

The weight of a body mass 20 lb is 20 lbf or 20×32, *i.e.* 640 pdl (taking g $= 32$). Hence the energy given to the body in raising it 10 ft is 20×10, *i.e.* 200 ft lbf or 640×10, *i.e.* 6 400 ft pdl. The potential energy of the body is therefore 200 ft lbf or 6 400 ft pdl (if g $= 32$).

4.6 ENERGY CONVERSION: WORK

Work is one of a number of English words which experiences a double usage as part of the language, one meaning being precise and scientific whilst the other in more general use is loose. The two meanings or shades of meaning do however share a common ground in that in the process known as work, energy of one kind or another is being used or, to be precise, converted. Newtonian physics envisaged a world and indeed a universe in which the quantity of energy is constant, only the nature varying. Thus Isaac Newton and his contemporaries believed implicitly that energy cannot be created or destroyed only converted from one form to another. With nuclear technology, which began with Einstein's theory of the equivalence of mass and energy, it has been shown that energy can in fact be created from mass. However, the Newtonian concept still applies to everyday mechanics and our definition of work, which must, of course, be precise, is based on energy conversion.

Definition
Work is done whenever energy is converted from one form to another. The amount of energy converted is taken as a measure of the work done and consequently the units of energy are also those of work.

It can be seen that since whenever a force moves its point of application energy is being used (converted), work is being done and this corroborates the definition of work found in the majority of other texts. It can also be seen that work is synonymous with *converted* energy.

4.7 RATE OF DOING WORK: POWER

Like the word 'work', power is another term widely used and often incorrectly so, one example being when it is used synonymously with 'energy'. Power and energy are not, of course, the same quantity; power is in fact a quantity directly derived from energy and therefore contains it in its definition.

Definition
Power is defined as the *rate of doing work* or, as follows from Section 4.6, the *rate of energy conversion*. The unit of power for any system is basically the unit of energy per unit time.

4.8 UNITS OF POWER

In the International System the unit of energy is the metre newton or joule and the unit of time is the second. Consequently, since the system is coherent, the unit of power is the metre newton per second (N m/s) or joule per second (J/s). *One joule/second* is given the special name the *watt* (abbreviated W), after the Scottish engineer James Watt who did considerable original work on projects involving mechanical and steam energy and power. The usual multiple units apply and we have the *kilowatt* (kW) and *megawatt* (MW), being 1 000 and 1 000 000 watts respectively, and also the sub-units *milliwatt* (mW) and *microwatt* (μW) being 10^{-3} (1/1 000) and 10^{-6} (1/1 000 000) watts respectively (*see* Appendix I).

The watt is now the accepted power unit for both electrical and mechanical engineers but it will no doubt be some time before the Imperial gravitational and non-coherent unit, the *horsepower*, falls from general use.

The basic power unit in the Imperial FPS Gravitational System is the *foot pound force* per *second* (ft lbf/s). Originally due to the work of James Watt and now accepted by definition, 550 ft lbf/s is taken as *one horsepower*, this being the supposed rate of working of a healthy dray horse (although it has been pointed out that it would not in fact do such a horse much good to work at this rate!). One horsepower (abbreviated h.p.) is, then, 550 ft lbf/s or 33 000 ft lbf/min and, as can be derived (*see* Chapter 10), is directly equal to 746 W. It is interesting to note that a 2 kW fire is using energy at approximately the same rate as three healthy dray horses according to Watt's original proposition!

The horsepower is not only non-absolute and incoherent but is rapidly becoming anachronous. However, it was sufficiently accepted to give rise to the metric horsepower known as *cheval vapeur* (CV) in France or Pferdestärke in Germany. Since the British version is to be replaced by the watt or its multiples the Continental versions will not be discussed further.

The power unit for the Imperial FPS Absolute System is the *foot poundal* per *second*, but it is not used in practice.

Example 4.4
Calculate the output power of a crane motor which raises a body of mass 1 ton a height of 30 ft in 30 s.

$$\text{Mass of the body} = 2\,240 \text{ lb}$$

$$\text{Weight of the body} = 2\,240 \text{ lbf}$$

Therefore

$$\text{force to raise body} = 2\,240 \text{ lbf}$$

Therefore

$$\text{work done in raising the body 30 ft} = 2\,240 \times 30 \text{ ft lbf}$$

This work is done in 30 s, therefore the rate of working or power

$$= \frac{2\,240 \times 30}{30} \text{ ft lbf/s}$$

$$= 2\,240 \text{ ft lbf/s}$$

$$= \frac{2\,240}{550} \text{ h.p.}$$

$$= 4\cdot07 \text{ h.p.}$$

$$= 4\cdot07 \times 746 \text{ W}$$

$$= 3\,036\cdot2 \text{ W}$$

$$= 3\cdot04 \text{ kW}$$

4.9   THE KILOWATT-HOUR

So far we have studied units of energy and work and then units of power. The unit of power, the kilowatt, gives rise to a further unit of energy which is important because at present it is the commercial unit of electrical energy.

Clearly, since power is the rate of energy conversion, then multiplying power by time gives the total energy used in that time. For example, any device working at the rate of 1 000 J/s for 10 s will convert 10 000 J during that period. The joule as a unit of energy is too small for commercial use and consequently the *kilowatt hour* (kWh) is chosen. To determine energy in kilowatt hours, the power in kilowatts is multiplied by the time in hours. Obviously since the kilowatt hour is a measure of energy there is a direct relationship between it and the joule (*see* Example 4.6).

Example 4.5
Calculate the daily cost of running a 1 000 W electric heater, four 250 W lamps and one 300 W television for 5 hours a day if the cost per unit (kWh) is 1·5p.

$$\text{Total power} = (1\,000 + (4 \times 250) + 300) \text{ W}$$

$$= 2\,300 \text{ W}$$

$$= 2\cdot3 \text{ kW}$$

$$\text{Total daily energy converted} = 2\cdot3 \times 5 \text{ kWh}$$

$$= 11\cdot5 \text{ kWh}$$

$$\text{and total cost} = 11\cdot5 \times 1\cdot5\text{p}$$

$$= 17\cdot25\text{p}$$

Example 4.6

Calculate the cost per joule if one unit of electrical energy costs 1·8p. One unit means one kilowatt hour.

$$1 \text{ kW} = 1\ 000 \text{ W}$$

$$= 1\ 000 \text{ J/s}$$

$$1 \text{ h} = 60 \times 60 \text{ s}$$

$$= 3\ 600 \text{ s}$$

Therefore

$$1 \text{ kWh} = 1\ 000 \times 3\ 600 \text{ J}$$

$$= 3·6 \times 10^6 \text{ J}$$

$$\text{Cost per kWh} = 1·8\text{p}$$

Therefore

$$\text{Cost per joule} = \frac{1·8}{3·6} \times 10^{-6}$$

$$= 0·5 \times 10^{-6}\text{p}$$

i.e. one half of one millionth of one penny per joule. This example shows that for the usual average cost of 1 kWh the joule is too small a unit for commercial use.

4.10 TORQUE

The final quantity to be discussed in this chapter has units *dimensionally* (*see* Chapter 10) the same as those of energy, but it is not in fact a direct measure of energy and requires multiplication by a non-dimensional (in the terms of reference of Chapter 10) quantity to become a direct measure. The quantity is *torque* and directly it is a measure of the *turning ability* of a force.

Torque is of interest where turning action is involved, *i.e.* rotating machinery such as motors, etc. It is measured by multiplying the peripheral force, *i.e.* the force acting along the periphery or circumference of a rotating shaft, by the radius of the shaft. Hence we may say that torque is a measure of *rotational moment* (to obtain the moment of a force about a particular point one multiplies the force by the distance of its point of application from the point). As can be seen the unit of torque will consist of the force unit and the unit of length, exactly as does the unit of energy. However, torque is not in fact a *direct* measure of energy but is an *indirect* measure inasmuch as it requires modification by a quantity not expressible in terms of the fundamental concepts mass, length and time, *i.e.* an *angle*, to become a direct measure. Dimensionally (*see* Chapter 10) the units of energy and torque are the same.

4.11   RELATIONSHIP BETWEEN TORQUE AND ENERGY

Figure 4.1 shows a shaft, radius *r* units of length, rotating at *N* rev/min, having a peripheral force, *i.e.* along a tangent to the circumference of the shaft, *T* units of force.

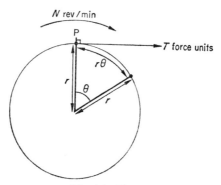

Fig. 4.1   Torque.

By definition the *torque* is *Tr* units. These units will be force–length units and are usually written as such: N m, pdl ft, lbf ft, etc. (compare energy units written N m, ft pdl, ft lbf, respectively).

Consider point P on the circumference. The distance moved by this point when the radius from P to the centre has passed through an angle θ is given by $r\theta$ units of length, if θ is expressed in radians. (The arc subtended by an angle θ between two radii of a circle radius *r* is given by (radius) × (angle), *i.e.* $r\theta$ provided that θ is expressed in radians. There are 2π radians in 360°, *i.e.* one radian is 57·3°.) Thus the energy converted by the tangential force *T* is given by

energy = (force) × (distance moved by point of application)

i.e.

$$energy = Tr\theta \tag{4.3}$$

and since torque = *Tr*, then energy = (torque) × θ.

The total energy for one complete revolution, *i.e.* 2π radians, is therefore given by

$$energy = 2\pi \times (torque) \tag{4.4}$$

so that, as can be seen, the torque exerted by the shaft is an indirect measure of the energy converted per revolution.

4.12 TORQUE AND OUTPUT POWER

Power is the rate of energy conversion. Hence for the case discussed in Section 4.11, in which the shaft is rotating at N rev/min, *i.e.* $N/60$ rev/s, the energy converted per second is given by

(energy converted per second) = (energy converted per revolution)
× (number of revolutions per second)

$$= 2\pi T \times \frac{N}{60} \qquad (4.5)$$

Thus the output power of a shaft rotating at N rev/min and having a torque T is given by $2\pi NT/60$ and the units are those of power, *i.e.* watts, ft lbf/s, h.p., etc.

PROBLEMS ON CHAPTER FOUR

(1) A cyclist rides a distance of $\frac{1}{4}$ mile along a horizontal road. The wind resistance is 4 lbf and the frictional resistance is 3 lbf. Calculate the work done by the cyclist. If the journey takes 1 min calculate the rate of working in horsepower.

(2) Calculate the time taken for a 5 h.p. lift motor to raise a lift weighing 2 240 lbf up a 50 ft shaft (neglect frictional losses due to winding pulleys, etc.).

(3) A scrap breaker of mass 1 000 lb is raised vertically to a height of 30 ft in 30 s and allowed to fall. Calculate (a) the output power of the motor in h.p., (b) the maximum potential energy of the breaker (in ft pdl), (c) the velocity of the breaker just before it hits the ground assuming all the potential energy is converted to kinetic energy (take $g = 32$ ft/s^2).

(4) Calculate the rate of working (in kW) if a 20 kg mass is raised vertically a distance of 100 m in 10 s (take $g = 9\cdot81$ m/s^2).

(5) Calculate the number of ft lbf in 1 kWh (1 ft $= 0\cdot304$ 8 m, 1 lb $= 0\cdot453$ 6 kg, $g = 9\cdot81$ m/s^2).

(6) A body is raised to a height of 10 m above the ground and allowed to fall. Assuming 100% energy conversion, calculate the velocity of the body just before it hits the ground (take $g = 9\cdot81$ m/s^2).

CHAPTER FIVE

Heat

5.1 THE NATURE OF HEAT, HISTORY OF RESEARCH

The nature of the phenomenon of heat has, surprisingly, puzzled scientists for many years and it is only comparatively recently (the latter half of the nineteenth century) that it has been explained in terms of quantities already recognised. Until 1840 it was believed that when a body changed its degree of heat, *i.e.* became colder or hotter, the change was due to the transfer of an invisible substance called 'caloric'. It was assumed that a body lost caloric on cooling and gained it on becoming hotter. Since weighing a body when hot and cold showed no difference in weight, the substance caloric was assumed to be weightless too. Crude though the theory seems in the light of present knowledge it did offer some explanation for the changes occurring when bodies are heated.

The caloric theory was abandoned owing to the research of several nineteenth century physicists, the most notable being Rumford and Joule. The first grave doubts as to the feasibility of the existence of caloric were consolidated by the work of Benjamin Rumford at the turn of the eighteenth century. Rumford was engaged by the Government of Bavaria in the boring of military cannon. At that time this was achieved by the boring of solid metal and he noticed the considerable amount of heat generated during this process. The caloric theory indicated that the production of heat in one body caused a loss in another as the caloric was transferred. The supporters of the theory suggested the caloric had been transferred from the chips caused by the boring, the chips then being 'too small to retain the caloric'.

Rumford countered this by a new boring with a blunt drill, immersing the whole thing in water. After some two and a half hours the water boiled and the number of chips amounted to only ten ounces in mass. The heat had been increased but the chips reduced. However, since there was not then a plausible alternative theory which could be practically demonstrated it was another forty or so years before the caloric theory was finally abandoned in favour of that proposed by James Prescott Joule. The experiment, which he performed in 1840 and which has become a classic of its kind, will be discussed more fully in Section 5.7. Joule related heat to mechanical energy and it is thus that the SI unit of the quantity of heat is the unit of energy, this being named in honour of the scientist.

5.2   QUANTITY AND DEGREE OF HEAT

The words 'heat' and 'temperature' are commonly misused in general use and the one quantity is frequently referred to by the name of the other. The distinction between temperature, which may roughly be termed as a measure of the degree or intensity of heat, and quantity of heat was first noted by Joseph Black (1728–1799). The fact that the two quantities are different is easily shown by considering, for example, a small quantity of molten copper at a temperature of 1 084°C (temperature scales are fully discussed below) and a pan of boiling water at a temperature of 100°C. The water will contain a far greater quantity of heat than the copper (the exact amounts depending on the respective masses and other considerations to be dealt with later), a fact easily demonstrated by observing the heat change of similar quantities of cold water when the two hot bodies are added to them. It is apparent that the temperature of a body is dependent on the heat quantity; how they are connected will be considered shortly.

Heat, in the precise sense, is now accepted as a form of energy and as such may be measured in units of energy and obtained by conversion from other energy forms (*i.e.* chemical, electrical, magnetic, mechanical, etc.). It may be transferred from place to place by conduction (through solids), convection (through gases) or radiation (via electromagnetic waves—*see* Chapter 7). Heat quantity is a measure of the total energy (kinetic and potential) of the molecules making up a substance. Temperature is a measure of the average kinetic energy of the molecules.

The product of temperature, mass (a measure of the total number of molecules) and a constant determined by the material (*i.e.* taking into consideration the *arrangement* of the molecules) gives a measure of the total heat quantity for a particular body. This is dealt with in more detail in succeeding sections.

5.3   TEMPERATURE MEASUREMENT

The temperature or 'hotness' of a body cannot be accurately measured by the human senses. This is clearly shown by considering the apparent degree of heat of water if checked first by a hand which is warm and then by a hand which is cold. The same water may appear hot or cold depending on whether the hand is cold or hot respectively. Temperature and temperature scales must be related to *reproducible* physical conditions. Accurate temperature measurement should be independent of the device or apparatus used in the measurement. Early temperature scales are due to the two main scientists Celsius and Fahrenheit. The Swedish astronomer Celsius devised a scale between the melting point of ice and the boiling point of water (1742), sub-dividing the interval between these so-called fixed points into 100 intervals called degrees. He called the lower point (melting point) 100 degrees and the upper point (boiling point) 0 degrees. Later these points were numerically reversed and melting point became 0 degrees and boiling point 100 degrees. The scale was then known as the *Centigrade* scale and one degree was written as 1°C. The points on this scale have now been redefined more precisely and the name has reverted to *Celsius* rather than Centigrade (this followed a decision of the 9th General Conference on Weights and Measures). The new definition of the lower point is discussed shortly.

Fahrenheit devised a scale (1712) between the melting point of ice, which he called 32 degrees and the normal body temperature, which he called 96 degrees (the latter temperature is now accepted as 98·4°F). Boiling point of water on the Fahrenheit scale was observed to be 212 degrees but the point was not taken as a fixed point in the original scale. Fahrenheit was a German physicist who first introduced the idea of using the expansion of mercury on heating as a means of measuring temperature, and it is on this latter point that the accuracy of using the original Celsius and Fahrenheit scales as a means of measuring temperature falls down, for, as was pointed out earlier in this section, temperature measurement should be independent of the means used. Any irregularities in the coefficients of thermal expansion of either the thermometric liquid (mercury, alcohol, etc.) or the container affect the accuracy of the measurement.

In 1848 Lord Kelvin introduced an absolute scale of temperature independent of the coefficients of thermal expansion of substances. For an ideal gas* at constant volume, pressure is directly proportional to temperature and pressure variations may be taken as a direct measure of temperature variations. If pressure and temperature could be continuously reduced in the same proportion a point would be reached at which the pressure, and therefore the temperature, would be zero. This point is called *absolute zero*. It is possible, using a

---

* An ideal gas is one in which the molecules are assumed to have negligible size and exert no force on neighbouring molecules. Collisions are assumed to be perfectly elastic. No gas is ideal but very close approximations may be obtained.

constant volume gas thermometer, to determine experimentally the value of any temperature on the Kelvin scale in terms of any other temperature on that scale, and in particular to determine how far above absolute zero the melting point of ice or any other fixed point lies.

Corrections still have to be made, of course, for the fact that the gas is not ideal and for the expansion of the gas container, but the scale may be considered to be absolute in so far as is presently possible. The Kelvin scale has been adopted, after suitable definitions were made, as the scale for use with the Système International. All three scales will be discussed in the following section.

The melting point of ice depends upon a number of variable quantities and the accuracy of reproduction of this point is open to question. A more accurately reproducible point (which can be reproduced to within 1/1 000 of one degree Celsius) is the *triple point* of water; this is the point at which water, water vapour and ice exist in stable equilibrium. It lies slightly above the melting point of ice at normal atmospheric pressure.

5.4   TEMPERATURE SCALES IN CURRENT USE

The absolute temperature scale accepted for use in the SI is the Kelvin scale. The triple point of water is *defined* as 273·16 Kelvins (abbreviated 273·16 K), the melting point of ice under normal atmospheric pressure as 273·15 K and the boiling point of water under normal conditions as 373·15 K. It will be noted that between the melting point of ice and the boiling point of water there are 100 intervals. The *Celsius* scale (often wrongly called Centigrade) has the melting point of ice at 0°C, the triple point of water at 0·01°C and the boiling point of water at 100°C, where the abbreviation ° signifies 'degree'.

As can be seen, the temperature interval Kelvin is equal to the temperature interval degree Celsius. Conversion of temperatures expressed in degrees Celsius to Kelvins or *vice versa* is achieved by addition or subtraction of 273·15 respectively.

The Fahrenheit scale has the melting point of ice as 32°F and the boiling point of water as 212°F. Conversion from Celsius to Fahrenheit or *vice versa* is achieved by applying the formulas shown in Fig. 5.1. If absolute zero is written on the Fahrenheit scale as zero degrees instead of $-459·67°F$ (*i.e.* the number of degrees Fahrenheit in 273·15°C), the scale is given the name *Rankine* (one degree Rankine is abbreviated 1°R)—*see* Fig. 5.1 and worked Examples 5.1, 5.2 and 5.3.

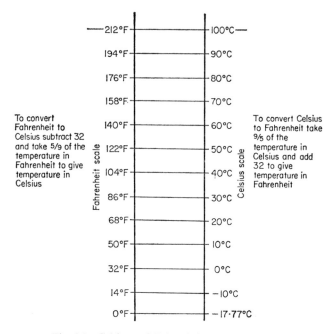

Fig. 5.1   Celsius and Fahrenheit temperature scales.

5.5   SOME WORKED EXAMPLES ON TEMPERATURE CONVERSION

Example 5.1

(a) Convert 92°C to degrees Fahrenheit. (b) Convert 73°F to degrees Celsius.

(a)
$$\frac{9}{5} \times 92 = \frac{828}{5} = 165 \cdot 6$$

Add 32 to give the Fahrenheit temperature, *i.e.* 197·6. Hence 92°C = 197·6°F.

(b) Subtract 32 from the Fahrenheit temperature, *i.e.*

$$73 - 32 = 41$$

$$\frac{5}{9} \times 41 = \frac{205}{9} = 22 \cdot 7$$

Hence 73°F = 22·7°C.

Example 5.2
Convert the temperatures given in Example 5.1 to Kelvins and degrees Rankine respectively.

(a)
$$92°C = (92 + 273·15) \text{ K}$$
$$= 364·15 \text{ K}$$

(b)
$$73°F = (73 + 459·67)° \text{ R}$$
$$= 532·67° \text{ R}$$

Example 5.3
(a) Convert 372 K to degrees Rankine. (b) Convert 622°R to Kelvins.

(a) Probably the easiest method is to convert to degrees Celsius, then to degrees Fahrenheit and finally to degrees Rankine.

$$372 \text{ K} = (372 - 273·15)°C$$
$$= 98·85°C$$
$$\frac{9}{5} \times 98·85 = \frac{889·65}{5}$$
$$= 177·93$$

Add 32 to give degrees Fahrenheit, *i.e.*

$$(177·93 + 32) = 209·93°F$$

Add 459·67 to give degrees Rankine, *i.e.*

$$(209·93 + 459·67) = 669·60°R$$

Hence 372K = 669·60°R
(b) Again the easiest way is to convert degrees Rankine to degrees Fahrenheit, then to degrees Celsius and finally to Kelvins.

$$622°R = (622 - 459·67)°F$$
$$= 162·33°F$$

Subtract 32, *i.e.*

$$(162·33 - 32) = 132·33$$
$$\frac{5}{9} \times 132·33 = \frac{661·65}{9}$$
$$= 73·51°C$$

Add 273·16 to give Kelvins, *i.e.*

$$(73·51 + 273·16) = 346·67 \text{ K}$$

Hence 622°R = 346·67 K

5.6   QUANTITY OF HEAT

As has been stated it is now recognised that heat is a form of energy and as such the quantity of heat may be expressed in energy units (*see* Chapter 4). Whilst this is now accepted practice in the International System, for a while earlier units of heat quantity will be encountered. These are the *calorie, kilocalorie, British Thermal Unit* and the *Centigrade* (Celsius) *heat unit.*

The *calorie* (abbreviated cal) is the amount of heat required to raise the temperature of one gramme of water from 14·5°C to 15·5°C (this is known as the 'fifteen degree calorie' and is not quite equivalent to the heat required to raise the temperature of one gramme of water by one degree Celsius anywhere on the scale since the physical properties of water change slightly with temperature).

The *kilocalorie* (kcal) is defined similar to the calorie, the difference being that the mass of water involved is one kilogramme. One kilocalorie is equal to 1 000 cal.

The *British Thermal Unit* (Btu) is the heat required to raise the temperature of one pound of water by one degree Fahrenheit. For precision it is taken as the average value for temperatures between the ice point and steam point on the Fahrenheit scale.

The *Centigrade heat unit* (Chu) is the amount of heat required to raise the temperature of one pound of water by one degree Celsius.

5.7   JOULE'S EXPERIMENT

Figure 5.2 shows the equipment used by Joule in 1840. This experiment was the first to show the connection between mechanical energy and heat. The potential energy of bodies was used to raise the temperature of water and a direct equivalence between mechanical energy and heat energy was obtained. The bodies were raised by a cord and pulley system without disturbing the water, a clutch pin was inserted and the falling bodies were allowed to turn a paddle in a container filled with water. The container had stationary vanes through which the paddle moved so that the water did not rotate as one body. Joule calculated the potential energy of the bodies (*see*

Chapter 4) and measured the initial temperature of the water and its mass and then the final temperature after the bodies had fallen. In this way he related energy units to heat quantity units. He realised that not all the potential energy was converted and made corrections for the cooling of the container, the kinetic energy of the bodies just before they reached ground level, the stretching of the cords, the friction of the pulleys and even the energy lost in the 'squeaking' of the pulleys as they rotated. His result differs by only 0·5% from results obtained in present day sophisticated apparatus.

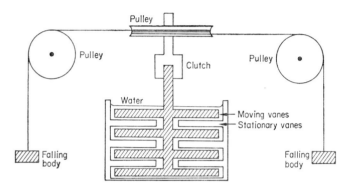

Fig. 5.2   Form of Joule's apparatus.

The connection between the then used heat units and energy units is called the *mechanical equivalent of heat* or *Joule's equivalent* and is equal to 4·19 J/cal.

The SI now uses the joule as the heat energy unit and if calories or other heat units are encountered it is necessary to apply Joule's equivalent to carry out conversion.

Example 5.4
Calculate the heat energy in joules required to raise the temperature of 15 g of water from 32°C to 56°C.

One calorie raises the temperature of one gramme of water by one degree Celsius. Hence, the heat required to raise the temperature of 15 g of water by (56 − 32), *i.e.* 24°C, is 15 × 24 cal.
There are 4·19 J/cal; hence this quantity of heat

$$= 15 \times 24 \times 4·19 \text{ J}$$

$$= 1\ 508·4 \text{ J}$$

(There is an approximation here since one calorie is defined using a specific part of the Celsius scale which is not involved in the example. However, the approximation is acceptable and will be used in succeeding examples without further comment.)

5.8   SPECIFIC HEAT CAPACITY AND SPECIFIC HEAT RATIO

It was indicated in Section 5.2 that the amount of heat absorbed or given up by a body undergoing a rise or fall in temperature depends upon the mass of the body, the temperature change and the nature of the material of the body.

As has been shown, the various heat units in use in systems other than the International System are defined by a specified change in temperature of a specified mass of water. It is to be expected that the effect on the temperature of a particular mass of water produced by a particular amount of heat will not necessarily be the same effect as that produced on the temperature of a body of the same mass by the same amount of heat if the body material is other than water. The effect depends not only upon the mass of the body but on its material since heat is a form of internal (or overall molecular) energy.

The amount of heat energy required to produce a one degree change in the temperature of a body of unit mass of any material is called the *specific heat capacity* of that material. The unit will be that of heat unit per mass unit per degree.

For the SI only, specific heat capacity of any material is defined as the amount of heat (in joules) required to change the temperature of 1 kg of the material by 1 K (or 1°C since they are equal). The unit is thus the joule per kilogramme per Kelvin (J/kg K). In practice the multiple kilojoule per kilogramme per Kelvin (kJ/kg K) is the preferred unit.

In other unit systems the specific heat capacity could be expressed in calories per gramme per degree Centigrade (CGS system), Btu per pound per degree Fahrenheit (FPS system), and so on.

Definition
The specific heat capacity of a material is the heat energy required to change the temperature of unit mass of the material by one degree. The symbol is s. For the SI the unit is the kilojoule per kilogramme per Kelvin (kJ/kg K).

It follows that if s is the amount of heat energy required to change the temperature of unit mass by one degree, then the amount of heat energy required to change the temperature of any mass m by δt degrees is given by

$$\text{heat energy} = m \times s \times \delta t \tag{5.1}$$

i.e. (heat energy) = (mass) × (specific heat capacity) × (change in temperature). The units of the quantities in this equation must of course be consistent, *i.e.* from the same system.

Because the calorie and British Thermal Unit are coherent units, *i.e.* defined in terms of the heat to give *unit* mass a temperature change of *one* degree, and because specific heat capacity is defined in terms of heat energy per unit mass per degree, the value of the specific heat capacity for a particular material expressed in calories per gramme per degree Celsius is numerically the same as when expressed in British Thermal Units per pound per degree Fahrenheit. This is shown below.

Example 5.5
Express a specific heat capacity of 0·2 cal/g °C in Btu/lb °F (1 lb = 453·6 g).

Now 1 cal changes the temperature of 1 g by 1°C, *i.e.* 1 cal changes the temperature of

$$\left(\frac{1}{453\cdot6}\right) \text{lb by} \left(\frac{9}{5}\right) \text{°F}$$

Thus 453·6 × 5/9 cal changes the temperature of 1 lb by 1°F, and since this is the definition of the Btu then 1 Btu = (453·6) × (5/9) cal, *i.e.*

1 Btu = (conversion factor from pound to gramme)
         × (conversion factor from °F to °C) × 1 cal

Writing Btu for cal, lb for g and °F for °C in the expression for specific heat, 0·2 cal/g °C is

$$0\cdot2\left(\frac{1}{453\cdot6} \times \frac{1}{5/9}\right) \text{Btu} \Big/ \left(\frac{1}{453\cdot6}\right) \text{lb} \left(\frac{1}{5/9}\right) \text{°F} \tag{5.2}$$

and clearly the conversion factors cancel leaving 0·2 Btu/lb °F, *i.e.* numerically the same as before.

(Had the Btu been defined, for example, as the heat energy to change the temperature of a mass of x lb by y°F the factors xy would have appeared in the numerator of eqn. (5.2) but not in the denominator, so that the resultant value for s in Btu/lb °F would not have been the same as cal/g °C but would have been multiplied by the factor xy.)

Thus if s is given in cal/g °C this value may also be taken as Btu/lb °F. This is *not* however true if s is expressed in J/kg K since the joule is not a coherent unit in terms of mass and temperature (it *is* a coherent unit in terms of force and distance), but a factor of 4 190 is involved, *i.e.* since

$$1 \text{ cal} = 4{\cdot}19 \text{ J}$$

and

$$1 \text{ g} = \frac{1}{1\ 000} \text{ kg}$$

$$1°C = 1 \text{ K}$$

then s cal/g °C is

$$s \times 4{\cdot}19 \text{ J} \Big/ \left(\frac{1}{1\ 000}\right) \text{ kg K}$$

i.e. 4 190 × s J/kg K.

When using s in problems its units should be carefully checked, especially at the present time, since the units of s as given are usually cal/g °C though, as will be discussed below, this is not always pointed out.

Example 5.6
Calculate the heat energy in joules required to change the temperature of a body of mass 10 kg from 10°C to 25°C if its specific heat capacity is 1 000 J/kg K.

The units of s do not have to be changed since heat energy in joules is required. The temperature scale employed is Celsius and not Kelvin but since the temperature *intervals* are the same the change in temperature as given, *i.e.* 15°C, is of course 15 K. Hence, using eqn. (5.1),

$$\text{heat energy} = 10 \times 1\ 000 \times 15$$
$$= 150\ 000 \text{ J}$$

Example 5.7
Calculate the specific heat capacity of a certain material if 300 cal are required to change the temperature of 50 g from 32°F to 95°F. Express in (a) cal/g °C, (b) Btu/lb °F, (c) J/kg K.

(a) A temperature change of (95 − 32)°F, *i.e.* 63°F is

$$\tfrac{5}{9} \times 63°C = 35°C$$

Hence, from eqn. (5.1),

$$300 = 50 \times s \times 35$$

and

$$s = \frac{300}{50 \times 35} \text{ cal/g }°C$$

$$= 0·171\ 4 \text{ cal/g }°C$$

(b) As was shown in Example 5.5, the value of s in cal/g °C is numerically that of s in Btu/lb °F. Hence

$$s = 0·171\ 4 \text{ Btu/lb }°F$$

(c) In finding s in J/kg K the heat energy in calories must be converted to joules and the mass in grammes to kilogrammes:

$$300 \text{ cal} = 300 \times 4·19 \text{ J}$$

$$50 \text{ g} = 0·05 \text{ kg}$$

and, since 35°C = 35 K, using eqn. (5.1),

$$300 \times 4·19 = 0·05 \times s \times 35$$

$$s = 718 \text{ J/kg K}$$

$$(i.e.\ 0·171\ 4 \times 4\ 190 \text{ J/kg K})$$

Specific heat ratio
There is considerable confusion between specific heat capacity and specific heat ratio owing to the method of definition of the calorie and the British Thermal Unit, but until the present time this confusion has not been over important except in the fact that a basic misunderstanding exists. With the advent of the SI and the formal adoption of one energy unit for use in all cases, the joule, which is not a coherent unit in terms of heat quantities (as are the calorie and British Thermal Unit), it is essential that the student should recognise the difference between these quantities and appreciate the source of the confusion.

Specific heat ratio is defined by the equation

specific heat ratio of any material

$$= \frac{\text{specific heat capacity of the material}}{\text{specific heat capacity of water}} \quad (5.3)$$

s being in the same units in the numerator and denominator.

Since the specific heat ratio is a ratio of two quantities having the same units, it is itself dimensionless, *i.e.* a pure number. Now the specific heat capacity of water, *i.e.* the quantity of heat required to change the temperature of unit mass by one degree, is *unity* when using cal/g °C or Btu/lb °F *by definition* (since the calorie and British Thermal Unit are *defined* in this way).

Consequently, when using these units in eqn. (5.3) the denominator is unity and the value of the specific heat ratio of a material is *numerically* the same as the value of the specific heat capacity of that material. They are not, of course, the same quantity since one is a ratio and dimensionless. Up to the present time the systems in common use have been those using the calorie and British Thermal Unit and consequently the confusion between the quantities has arisen even to the point where basic texts use the term 'specific heat' (which is meaningless without further qualification). The reason that it now becomes important to recognise this confusion and avoid such misunderstanding in the future is that with the new SI the specific heat ratio does *not* have the same value as the specific heat capacity for a particular material since the specific heat capacity for water expressed in SI units is *not* unity.

Example 5.8

Find the specific heat ratio of the material of Example 5.7 using the specific heat capacity expressed in the three kinds of units.

From Example 5.7 for the material given

$$s = 0{\cdot}171\ 4\ \text{cal/g °C}$$
$$= 0{\cdot}171\ 4\ \text{Btu/lb °F}$$
$$= 718\ \text{J/kg K}$$

For water

$$s = 1\ \text{cal/g °C}$$
$$= 1\ \text{Btu/lb °F}$$
$$= 4\ 190\ \text{J/kg K}$$

Hence specific heat ratio is

$$\frac{0{\cdot}171\ 4\ \text{cal/g °C}}{1\ \text{cal/g °C}} = 0{\cdot}171\ 4$$

or

$$\frac{0{\cdot}171\ 4\ \text{Btu/lb °F}}{1\ \text{Btu/lb °F}} = 0{\cdot}171\ 4$$

or

$$\frac{718\ \text{J/kg K}}{4\ 190\ \text{J/kg K}} = 0{\cdot}171\ 4$$

As can be seen, the specific heat ratio of a material is the same whichever units are used (which is to be expected since it is dimension-less) and it has the same numerical value as specific heat capacity when cal/g °C and Btu/lb °F units are involved, but *not* when J/kg K units are involved.

It will be some time before all references to 'specific heat' and tables giving values of 'specific heat' go from the bookshelves and it is therefore important firstly to realise that such tables actually give values of *specific heat capacity*, usually being expressed in cal/g °C, and secondly to be able to convert to the more correct SI units.

Example 5.9
Calculate the heat required to change the temperature of 10 lb of a certain material from 20°C to 40°C. The specific heat ratio is 0·21. Obtain the answer in (a) British Thermal Units, (b) calories, (c) joules.

The fact that the specific heat *ratio* is 0·21 tells us that the specific heat *capacity* is 0·21 cal/g °C or 0·21 Btu/lb °F or 4 190 × 0·21 J/kg K.

(a) The temperature change is 20°C, *i.e.* 36°F. Hence from eqn. (5.1) heat required

$$= 10 \times 0.21 \times 36 \text{ Btu}$$

$$= 75.6 \text{ Btu}$$

(b) The mass is 10 × 453·6 g. Hence from eqn. (5.1) heat required

$$= 10 \times 453.6 \times 0.21 \times 20$$

$$= 19\,050 \text{ cal}$$

This answer could be obtained directly from the overall conversion equation 1 Btu = 252 cal, obtained in Example 5.5 and from the answer in (a) above.

The student should not infer from the fact that *s* expressed in cal/g °C units has the same value as *s* expressed in Btu/lb °F units that the British Thermal Unit is the same as the calorie. This example and that of Example 5.5 shows clearly that this is not so.

(c) Using eqn. (5.1) heat required

$$= \left(\frac{10 \times 453.6}{1\,000}\right) \times (4\,190 \times 0.21) \times 20$$

$$= 79\,820 \text{ J}$$

Note that care is taken in each case to use consistent units, *i.e.* from the same system.

5.9   LATENT HEAT

When a body changes its state from solid to liquid or liquid to vapour it is found that extra heat is required to complete the conversion. Whilst the conversion takes place, *i.e.* solid to liquid or liquid to vapour, though heat energy is absorbed there is *no* change in temperature. The energy is in fact required for the molecular changes taking place within the material but does not increase the level of molecular kinetic energy (*i.e.* does not change the temperature). On reversing the conversion process, *i.e.* vapour to liquid or liquid to solid, the same amount of heat taken in during the original conversion is given out, the material staying at the same temperature (melting point or freezing point respectively) all the while.

This heat quantity is given the name *latent* (hidden) *heat* and is further qualified by attaching the name of the conversion, *i.e. latent heat of evaporation*—taken in for liquid to vapour conversion, given out for vapour to liquid conversion—and *latent heat of fusion*—taken in for solid to liquid conversion, given out for liquid to solid conversion.

The total latent heat depends on the type of material and its mass. Consequently, latent heat is measured in heat energy units per unit mass (cal/g, Btu/lb and, most important of all, J/kg for the SI). In computing the heat energy required to raise the temperature of a body and to change its state to the next one up, the appropriate latent heat must be added to the quantity determined by eqn. (5.1). Similarly, in determining the heat given out when the temperature of a body drops, if the body passes through a conversion state en route the latent heat must be added to the heat quantity determined by eqn. (5.1).

Example 5.10
An electric furnace smelts 85 kg of tin every hour from an initial temperature of 15°C. Find the heat energy required per hour assuming the melting point of tin to be 235°C and the latent heat of fusion to be 55 000 J/kg (specific heat capacity of tin = 235 J/kg K). Use this answer to determine the furnace power (in kW).

The total heat energy is equal to the heat energy required to raise the temperature of 85 kg of tin from 15°C to 235°C plus the latent heat required to convert 85 kg of solid tin to liquid tin at 235°C (the melting point). Therefore,

$$\text{total heat} = 85 \times 235 \times (235 - 15) + 85 \times 55\,000 \text{ J}$$

$$= 9\,070\,000 \text{ J}$$

This energy is used every hour, *i.e.* every 3 600 s. Therefore,

$$\text{power} = \frac{9\ 070\ 000}{3\ 600} \text{ J/s or W}$$

$$= \frac{9\ 070}{3\ 600} \text{ kW}$$

$$= 2 \cdot 52 \text{ kW}$$

PROBLEMS ON CHAPTER FIVE

(1) Determine (a) the point on the Fahrenheit scale corresponding to 45°C, (b) the point on the Celsius scale corresponding to 59°F, (c) the point on the Kelvin scale corresponding to 59°F, (d) the point on the Rankine scale corresponding to 45°C.

(2) Determine the number of (a) calories required to change the temperature of 150 g of water from 32°C to 99°C, (b) joules required to change the temperature of 25 g of silver from 15°C to 35°C (specific heat capacity of silver = 0·055 6 cal/g °C), (c) British Thermal Units required to change the temperature of 25 g of alcohol by 25°F (specific heat ratio of alcohol = 0·547).

(3) A 1 kW heater takes 1 min 40 s to raise the temperature of 100 kg of a certain material by 20°C. Calculate the specific heat capacity of the material in SI units.

(4) Calculate the time taken for a 3 kW heater to melt 1 000 g of tin which is initially at 20°C (for tin $s = 0 \cdot 235$ kJ/kg K; latent heat of fusion = 55 000 J/kg; melting point = 235°C).

(5) Using the relevant information given in Problem 4 calculate the specific heat ratio of tin.

CHAPTER SIX

Light

6.1 DISCUSSION

With the invention by Edison and Swan of the incandescent lamp in 1879 the study of light took a step forward from an academic and somewhat unprecise subject interesting only physicists towards being a part of the ever expanding complex technology of modern times. Even at that time, however, although the 'electric light' was hailed with tremendous enthusiasm on all sides, the all pervading approach was a matter of the indiscriminate placing of a lamp wherever light was needed and there was little if any discussion of the type of lamp, the number, the output, the siting and the illumination required taking into consideration the functional aspects of the demand. Over the years, with further discoveries, including the use of tungsten filaments, coiled coil filaments, gas discharge lamps including the fluorescent tube, the science of light has become an important technology in its own right and illumination engineering plays a large and important part in every aspect of modern living. A technology demands precision of communication and thus a system of units, the standards of which are practical, understandable and easily reproduced.

6.2 THE NATURE OF LIGHT, HISTORY OF RESEARCH

As with many concepts in engineering, *e.g.* force (Chapter 3), energy (Chapter 4), electric charge (Chapter 7), light is an intangible quantity, in some respects even more so than other quantities since, although we recognise it by its effects, the limits of our recognition are personal and depend on human sensory organs which are as individual as people themselves. There can therefore be no set, unalterable, independent standard as with, for example, the units of length or mass, which are tangible quantities, and it is for this reason that for many years the science of light has been unprecise. Modern illumination engineering appears to have made as precise a job as possible under difficult circumstances, and has gone much of the way towards the goals outlined in Section 6.1.

Light itself can be considered to be a remotely induced effect, *i.e.* something which is experienced and which is due to a cause or source though there is no tangible medium between the source and the place where the effect is felt. To clarify this point consider the transmission of mechanical energy from the engine of a car to the drive wheels or of sound (which is a form of mechanical energy) from the source to the ear. In both cases the medium between cause and effect is tangible—in the case of the car, the pistons, the drive shaft, the gear wheels, etc., or in the case of sound, the air or other medium between the sound source and the listener—and, even more important, if the medium is removed the transmission of energy ceases. Light, on the other hand, needs no medium (or at least no medium of which present technology is aware).

This point proved a serious drawback to early scientists for whilst the early theory of Christiaan Huygens (1629–1695), which explained light radiation in terms of a vibration (like sound), went a considerable way towards explaining many of the then known effects of light, it required a medium. The sound of an electric bell in an inverted glass jar dies away as the air is evacuated but the light from a source similarly situated does not. Huygens suggested there was a medium, a tasteless, odourless, weightless material called ether, which could not be seen, demonstrated or measured. Modern science has demonstrated that if it exists its specific gravity must be 16×10^9 tons per cubic inch so it would seem Huygens was wrong!

Sir Isaac Newton, whose work on another intangible quantity, force, was discussed in Chapter 3, suggested light radiation was composed of a series of very tiny particles and though this overcame the medium problem and explained certain of the effects not explained by Huygens' theory, it did not explain a great many of the effects which were explained by Huygens and so popular support went to Huygens despite the ether.

Modern science still does not explain light radiation, at least not to the satisfaction of a purist; it does, however, go a long way to explaining many of the observed effects and to a considerable degree absorbs the best parts of both Huygens' and Newton's theories.

6.3 ELECTROMAGNETIC RADIATION

It is sometimes said that modern science has not explained fundamentals any better or more clearly than early science—it has merely reduced the number of inexplicable phenomena by relating them. The situation under discussion is a case in point.

It will be recalled that the transmission of light does not require a medium; in this respect it is not alone. The effects of an electric field or of a magnetic field are felt whether or not a medium is present. It is true that certain kinds of medium considerably strengthen these effects, but in both cases the media are judged by comparison with a vacuum. This is discussed in some detail under 'Field Quantities' (Chapter 8).

It has been found that whenever an electric field and magnetic field act together in a particular way *radiation* occurs and energy is transmitted. The nature and physical effects of the electromagnetic radiation depend upon the characteristics of the fields causing it and can vary over a wide range called a *spectrum*. Certain parts of this range can be displayed by tangible means, which considerably aids our understanding of the remainder of the range.

The radiation takes place in what is called a *wave motion* and because we cannot say of what—since the concept of the ether is not acceptable—the intangible is explained in terms of the tangible, *i.e.* wave motion is explained using an analogy which involves a physical medium having mass, weight and shape.

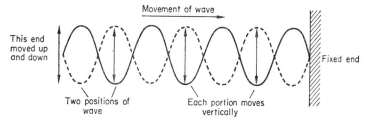

Fig. 6.1   Formation of a wave using a rope.

Consider a rope tied at one end whilst the other end is moved up and down in a vertical plane. The reader can satisfy himself as to what occurs by carrying out a practical demonstration. A disturbance moves down the rope and the rope takes up the shape of a wave which is moving down the rope towards the fixed end. In detail, what happens is this: each part of the rope remains the same distance from the fixed end but moves up and down in a vertical plane. At any instant the relative positions of each part are such that a wave shape is formed as in Fig. 6.1. Shortly after this instant of examination, each individual part of the rope has changed its vertical position slightly (whilst maintaining its horizontal position, *i.e.* the distance from the fixed end is the same) and the rope now forms a new wave shape of the *same* form as before but shifted towards the fixed end. So whilst each individual part of the rope only moves in a vertical plane, the wave travels in a horizontal plane from the vibrating end to the fixed end. Figure 6.1 helps clarify this point.

A wave of this nature, which it is believed is perfectly analogous to the method in which electromagnetic radiation is propagated, has certain defined physical characteristics which are important. The velocity with which the wave travels down the rope (*i.e.* through the

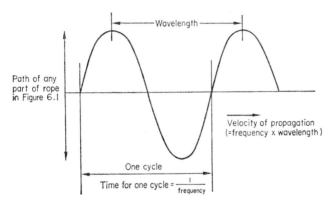

Fig. 6.2   Properties of a wave.

medium) is called the *velocity* of *propagation*, the symbol (for electromagnetic waves) is c and the quantity is measured in metres per second. The distance between any two similar points on the wave, *i.e.* between peaks or troughs, etc., is called the *wavelength*, symbol λ (lambda), and is measured in metres or sub-units of metres (*see* later). The complete excursion of the moving end of the rope from peak (or top vertical position) to trough (or bottom vertical position) and back to peak is called one *cycle* and the number of cycles executed per second is called the *frequency*, symbol f. The unit *one cycle* per *second* is given the special name *hertz*, abbreviated Hz. Since one complete cycle produces one complete wave, a complete wave may also be termed a cycle (*see* Fig. 6.2).

If one considers the wave front of a travelling wave such as the one discussed then if f waves are being transmitted per second and each wave is λ metres long the velocity of propagation of the wave front, *i.e.* the distance moved per second by the wave front, is the product of the number of waves per second, f, and the length of

TABLE 6.1

Electromagnetic radiation spectrum
(limits are approximate only)

Type	Frequency	Wavelength	Characteristics
Radio waves	3 kHZ–300 GHZ	100 km–1 mm	All radio frequency (rf) waves generate electrical energy in receiver aerials
including:			
Very low frequency VLF	3 kHz–30 kHz	100 km–10 km	
Low frequency LF	30 kHz–300 kHz	10 km–1 km	Long wave ⎫ broad-
Medium frequency MF	300 kHz–3 MHz	1 km–100 m	Medium wave ⎬ casting
High frequency HF	3 MHz–30 MHz	100 m–10 m	Short wave ⎭
Very high frequency VHF	30 MHz–300 MHz	10 m–1 m	Used in TV communication
Ultra high frequency UHF	300 MHz–3 GHz	1 m–10 cm	Used in TV, radar
Super high frequency SHF	3 GHz–30 GHz	10 cm–1 cm	Used in radar
Extremely high frequency EHF	30 GHz–300 GHz	1 cm–1 mm	Experimental
Radiant heat Infra-red	3×10^{11} Hz–3×10^{14} Hz	10^{-1} cm–10^{-4} cm	Heats surfaces and causes a small amount of chemical activity
Light waves			Produces a visual sensation in the brain; is chemically active in certain tissues

including:

	Frequency	Wavelength	Properties
Red	0.4×10^{15} Hz–0.46×10^{15} Hz	7.5×10^{-5} cm–6.5×10^{-5} cm	
Orange	0.46×10^{15} Hz–0.51×10^{15} Hz	6.5×10^{-5} cm–5.9×10^{-5} cm	
Yellow	0.51×10^{15} Hz–0.56×10^{15} Hz	5.9×10^{-5} cm–5.3×10^{-5} cm	
Green	0.56×10^{15} Hz–0.61×10^{15} Hz	5.3×10^{-5} cm–4.9×10^{-5} cm	
Blue	0.61×10^{15} Hz–0.71×10^{15} Hz	4.9×10^{-5} cm–4.2×10^{-5} cm	
Indigo	0.71×10^{15} Hz–0.73×10^{15} Hz	4.2×10^{-5} cm–4.1×10^{-5} cm	
Violet	0.73×10^{15} Hz–0.77×10^{15} Hz	4.1×10^{-5} cm–3.9×10^{-5} cm	
Ultra-violet radiation	10^{15} Hz–7.5×10^{16} Hz	3×10^{-5} cm–4×10^{-7} cm	Causes considerable chemical action in animal tissue (tanning of skin); intensive radiation can cause damage (snow blindness); causes fluorescence[a]
X-radiation (X-rays)	7.5×10^{16} Hz–10^{20} Hz	4×10^{-7} cm–3×10^{-10} cm	Pass through solids; can destroy living tissue, causes intensive fluorescence[a]
Gamma radiation (γ-rays)	10^{20} Hz upwards	3×10^{-10} cm upwards	Same properties as X-rays but more pronounced
Cosmic radiation (cosmic rays)	As gamma radiation		Originates in outer space; can pierce up to 15 ft of solids; may retard the growth of living creatures

[a] Some substances emit light when exposed to certain radiation. The process is called *fluorescence*.

each wave, λ, *i.e.*

$$c = f\lambda \tag{6.1}$$

or in words

velocity of propagation = (frequency) × (wavelength)—*see* Fig. 6.3

The velocity of propagation for all electromagnetic waves, and thus of light, is constant at 3×10^8 m/s (about 186 000 mile/s).

The nature of electromagnetic waves and the effects produced by them depend upon their frequency and wavelength. From eqn. (6.1) it follows that, since c is a constant, frequency is inversely proportional to wavelength. As was stated above, their nature depends upon the characteristics of the electric and magnetic fields causing

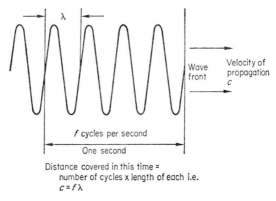

Fig. 6.3   Relationship between velocity of propagation and frequency and wavelength.

them; as can be inferred the frequency and thus the wavelength are determined by the characteristics of the cause. A breakdown of the nature of electromagnetic waves by frequency is given in Table 6.1.

6.4   THE CHARACTERISTICS OF LIGHT RADIATION AND LIGHT

The quantities 'light' and 'light radiation' are in fact two distinct quantities though they are often loosely referred to as the same thing. In fact, light radiation is a transmission of energy via electromagnetic waves and constitutes a very small part of the entire electromagnetic spectrum. Light, on the other hand, is a sensation produced by visual organic apparatus. Light radiation and light can be considered to be 'cause' and 'effect'.

The mechanism of the effect, briefly, is that energy received by radiation causes chemical reactions within the living tissues of the sensory organ, the eye, which in turn set up minute electric currents in the nerve fibres of the brain. These currents in turn, when passed to the optic lobes evoke the sensation we refer to as 'light'. As was stated in Section 6.2, the sensory organs of animals differ considerably and it is therefore not possible to set up an accurate system of measurement using only the 'effect' (*i.e.* light) and reference must be made instead to the 'cause' (*i.e.* light radiation) which is a measurable *physical* phenomenon. For the purpose of technology the immeasurable effect is closely identified with the measurable cause in the setting up of a precise system of measurement.

6.5 LIMITING WAVELENGTHS: THE LIGHT SPECTRUM

The wavelengths of radiation within the light band are so small that the metre is too large a unit for convenience. Instead, a small sub-unit of the metre, the *nanometre* (symbol nm), is used. *One nanometre* is one-thousandth of one-millionth of a *metre*, *i.e.* 1×10^{-9} m (or, alternatively, one-millionth of a millimetre, *i.e.* 1×10^{-6} mm). Radiations lying outside the limits 380 nm and 760 nm have no visual effect on the human eye. The longer wavelength is the limit of the *red* end of the light spectrum, the shorter wavelength the limit of the *violet* end. Between these limits lie the colour sensations *red/orange/yellow/green/blue/indigo/violet*. White light is composed of all these radiations and by suitable arrangement of glass prisms or other apparatus it can be split into its individual components.

6.6 A SYSTEM OF MEASUREMENT: LUMINOUS FLUX

In Section 6.4 it was stated that since light has an immeasurable effect on human sensory organs which differ widely, in formulating a system of measurement the measurable 'cause', *i.e.* light radiation, is used. A discussion follows of how this may be done.

As has been shown, light is a form of electromagnetic radiation and this, in turn, is a transmission of energy. A first approach to setting up measurable quantities lies in consideration of the rate of energy transmitted per unit time by this radiation. In comparing various radiations, rate of energy per unit time or power transmitted is not sufficient alone and the receptive area must be taken into account. In other words, for general comparison of electromagnetic radiation the power per unit area of the radiation is considered. The SI units of power and area are *watts* and *square metres* respectively.

For light radiation it is found that a straight comparison of power/ area levels does not give a true picture since the human eye responds differently to light radiations depending upon their wavelengths (*see* Fig. 6.4). This graph plots the power per unit area to give the same effect on the eye against the wavelength of the radiation. Before the inferences drawn from the graph are discussed, consider first how the graph is obtained. The words 'to give the same effect' are used. Now it has already been stated that the sensitivity of the eye varies from person to person so these words and the method of comparison must be discused in more detail.

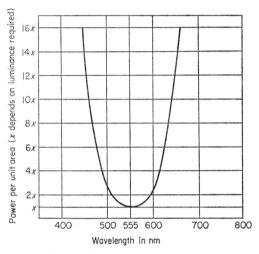

Fig. 6.4   Power/area vs wavelength.

To make a comparison a surface is irradiated with light of a particular frequency and wavelength and the objective brightness is noted by an observer (objective brightness will be discussed in more detail later). A second surface whose brightness or reflective properties are adjustable is then similarly irradiated and the surface adjusted until the same brightness is achieved as on the first surface *as judged by the same observer*. The conditions of the second surface are maintained whilst light of variable wavelength and power/area levels is used to irradiate the first surface. For each wavelength the power/ area level is adjusted for the same brightness as on the second surface and a series of values of power/area to give the same brightness is obtained. The curve is as shown. It is found that the maximum effect for minimum power/area is achieved by light of wavelength 555 nm and as far as can be ascertained this is true for the human eye in general. Since the same observer makes the comparison the fact that the *actual level of brightness* differs from person to person, depending upon age, physical condition, race, etc., does not enter into it. To clarify this consider three observers X, Y and Z performing the experiment. All three will find the same *relative* levels of power per unit area for a particular brightness compared with the power per unit area for light of wavelength 555 nm to be the same, although observer X may find that a radiation of a particular power/area and wavelength gives him a greater sense of brightness than the same

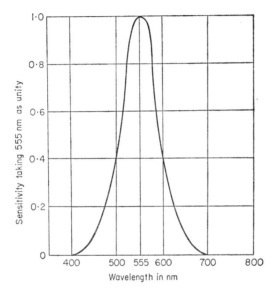

Fig. 6.5   Sensitivity vs wavelength.

levels give observer Y or Z. Since a comparison is being made, *actual* levels do not matter. Such methods are the basis of all light measurements.

So, as can be seen, the human eye is most sensitive to light radiation of wavelength 555 nm. Observations have shown that midday sunlight radiations radiate most power at this wavelength, which is, of course, why direct sunlight is so bright to all functioning human eyes whatever their condition.

The *sensitivity of the eye* to a particular light radiation is a measure of the relative effect of the light radiation compared with light radiated at the standard wavelength. A graph of sensitivity against wavelength is shown in Fig. 6.5. As can be seen, the sensitivity is *unity* for radiation at the optimum wavelength and less than this for all other wavelengths. This curve is called the *International relative luminosity curve.*

In addition to the power radiated per unit area for a particular radiation, we must therefore also take into consideration its wavelength and the relative effect it has on the eye compared with the optimum or standard radiation of wavelength 555 nm.

Multiplying the power per unit area of a particular radiation by the sensitivity factor for the appropriate wavelength gives a measure of the radiation in terms of its effect; further, it is a reproducible measure and does not rely on *individual* sensitivity.

The light energy transmitted by electromagnetic radiation may be considered to 'flow' from source to receiver; it is given the name *luminous flux*, symbol Φ. A measure of luminous flux may be obtained by multiplying together the power/area of the radiation and the sensitivity factor at the appropriate wavelength.

If the power is measured in watts per square metre, since the sensitivity factor is a ratio and thus dimensionless the unit of luminous flux is a watt/(metre)^2.

In this context the unit is given the name *light-watt*. In practice the light-watt is too large and a smaller unit, the *lumen* (abbreviated lm), where 630 lm = 1 light-watt, is used. Hence

$$\text{luminous flux} = \frac{\text{power}}{\text{area}} \times (\text{sensitivity factor}) \quad \text{light-watts}$$

and

$$\text{luminous flux} = \frac{\text{power}}{\text{area}} \times (\text{sensitivity factor}) \times 630 \quad \text{lm} \qquad (6.2)$$

Definition
Luminous flux is light energy evaluated in terms of the spectral sensitivity of the eye. Its symbol is Φ (phi) and the unit is the lumen, abbreviated lm (the lumen dimensionally is in energy units per unit area).

6.7 LUMINOUS INTENSITY

Energy transmitted by electromagnetic radiation is a vector quantity, *i.e.* it has both magnitude *and* direction. Therefore, in comparing the light radiating ability of sources it is not sufficient to consider magnitude alone. A source of light emits radiation in many directions simultaneously and these are not confined to a two dimensional plane, *i.e.* one having height and width only. Hence in considering direction a three dimensional or 'solid' measure must be used. The measure used is the *solid angle* unit, the *steradian*, abbreviated sterad. To appreciate the meaning of this unit of solid geometry consider a sphere of any radius *r*. Now consider a cone within the sphere with apex at the centre. The surface of the cone contains a portion of the volume of the sphere and cuts off a portion of the surface area of the sphere. When the apex angle of the cone is such that this portion of surface area equals r^2 square units, the solid angle contained by the cone surface at the apex is called one steradian. A two-dimensional drawing of this is shown in Fig. 6.6.

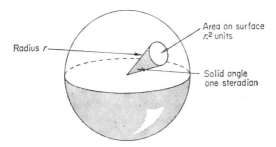

Radius *r*

Area on surface r^2 units

Solid angle one steradian

Fig. 6.6   The steradian unit of solid angle.

In considering the light radiation emitted by a source, then the radiation contained within a solid angle is used as an appropriate measure. The steradian is in fact a fairly large solid angle and in considering the luminous flux per steradian the assumption that the distribution is uniform throughout is not really valid. To make the assumption valid a very small solid angle is chosen and for radiation in a single direction the solid angle is taken to the limits of smallness.

The *luminous intensity* of a light source in a particular direction is defined as the light radiating ability in that direction. It is measured in units of luminous flux per solid angle, *i.e.* lumens per steradian. As stated above, for a specific direction (an extremely difficult situation to arrange for a light source in practice) the solid angle involved is very small indeed. The lumen per steradian is given a special name the *candela* (pronounced can-deela), abbreviated cd. The symbol for luminous intensity is I.

[This unit is in fact considered to be one of the six basic units of the Système International though as can be inferred from the previous discussion it is derived in terms of concepts already discussed (energy, area, etc.).]

Although the candela is equal to one lumen per steradian (1 lm/sterad), it is not defined in this way; to appreciate how it is defined it is first necessary to discuss the quantity luminance.

Definition
Luminous intensity of a source in a particular direction is the light radiating ability in that direction. The symbol is I and the unit is the lumen per steradian or *candela* (abbreviated cd).

6.8 LUMINANCE

The word 'brightness' as used generally describes a sensation which is variable from person to person and further depends to some degree on the physical situation of the source. Two equally bright sources—as determined using means other than the human eye—may not appear equally bright if one of them is surrounded by a light background and the other by a dark background. The source with the dark background will appear brighter to the eye. This leads to the necessity of further qualifying the word 'brightness': *subjective* brightness is the name given to the quality of brightness when the background is taken into consideration; *objective* brightness is the brightness regardless of surroundings and being precise and able to be evaluated without reference to its effect on human sensory organs is the usual meaning of 'brightness' in illumination engineering.

In considering the brightness of a source of light, whether it is a primary source (the sun, lamps, etc.) or a secondary source (a source by reflection, *e.g.* mirrors, polished or otherwise, reflective surfaces), it is necessary to take into consideration the area or *apparent* area of the source, and in comparing two or more sources it is more accurate from the point of view of getting a true comparison to consider the objective brightness per unit area of each source rather than of the source as a whole.

Objective brightness per unit area is given the special name *luminance,* symbol L, and it is measured in units of luminous intensity per unit area, *i.e. candelas* per *square metre*. Since luminance is dependent on luminous intensity, which has direction associated with it, luminance is also given as in a particular direction.

Definition
Luminance is the name given to the objective brightness of a source in a particular direction and is the luminous intensity in that direction per unit of apparent area of the light source. The symbol is L and the unit is the *candela* per *square metre* (cd/m^2).

It will be noticed that the words 'apparent area' appear in the definition. It is important to realise why this is so. The actual physical size of a light source is irrelevant as far as the eye is concerned; what is relevant is how the size appears to the observer or sensing apparatus. This is indicated more clearly diagrammatically in Fig. 6.7.

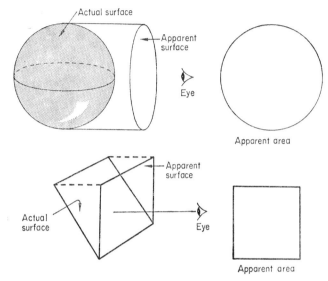

Fig. 6.7 Apparent area.

6.9 THE DEFINITION OF THE UNIT OF LUMINOUS INTENSITY

As stated in Section 6.7, the unit of luminous intensity in the International System is the candela and this unit is taken as one of the fundamental units of the system.

The candela is equal to one lumen per steradian, where the lumen is the unit of luminous flux as determined by eqn. (6.2). In order that the actual physical size of the unit should be demonstrated for practical use, it is necessary to select a particular light source which is invariable and easily reproducible. Standards of this nature have been altered over the years from a wax candle of determinate dimensions, which was somewhat unprecise by modern standards of technology, to certain kinds of lamps and finally to the light emitted by platinum as it solidifies (at 1 773°C). This light standard was adopted in 1948, sometime before the SI came into being and in fact before the candela was adopted. The standard was, however, so accurately and invariably reproducible that it remained. The luminous intensity of such a source is in fact 60 cd and accordingly the candela is defined using this standard.

Definition
A radiator (of light energy) has a luminous intensity of one candela if its luminance is 1/60 of that of a radiator maintained at the temperature of solidification of platinum.
Note that the radiators are compared by their luminance, *i.e.* their luminous intensity per unit of apparent area. In order to make a comparison the light is emitted through an aperture of known dimensions.

6.10 ILLUMINATION

So far the quantities considered have been directly concerned with the source of light energy. Some of the light energy radiated from a source is dissipated in space, but the majority falls upon a surface of one kind or another. To obtain some measure of the light energy reaching a surface the luminous flux incident upon the surface is divided by the area of the surface to give an average value of the flux per area. This quantity is called *illumination*, symbol E. The SI unit is the *lumen* per *square metre*. One lumen/square metre is called one *lux*, abbreviated lx.

Definition
Illumination is the amount of luminous flux incident on a surface per unit area. The symbol is E and the unit the *lumen* per *square metre*. One lumen/square metre is called one *lux*, abbreviated lx.

TABLE 6.2
Summary of light quantities

Quantity	Symbol	Relationships, units, remarks
Luminous flux	Φ	lumen (lm), may be derived from eqn. (6.2)
Luminous intensity	I	candela (cd), 1 cd = 1 lm/sterad
Luminance	L	candela/(metre)^2, (cd/m^2), luminous intensity per unit of apparent area of source
Illumination	E	lumen/(metre)^2, (lm/m^2), incident flux per unit area

The name formerly given to the unit one lumen per square *foot* is called one *foot-candle*, but this is not of course the unit of the International System.

6.11 SOME WORKED EXAMPLES ON CHAPTER 6

Example 6.1
The frequency of a certain electromagnetic radiation is 6×10^{14} Hz. Find the wavelength and state the type of radiation.

Frequency, velocity and wavelength are connected by the equation

$$\text{velocity} = (\text{frequency}) \times (\text{wavelength})$$

hence

$$3 \times 10^8 = 6 \times 10^{14} \times \lambda$$

Therefore

$$\lambda = \frac{3 \times 10^8}{6 \times 10^{14}}$$

$$= 500 \times 10^{-9} \text{ m}$$

$$= 500 \text{ nm}$$

The radiation lies in the visible light band, limits 380 nm to 760 nm.

Example 6.2
Calculate the power per unit area of the radiation of Example 6.1 required to give the same objective brightness as a radiation of wavelength 550 nm radiating x watts per square metre if the relative luminosity factor is 0·4. Hence determine the luminous flux of the radiation of Example 6.1 (assume the radiators are identical).

The value of 0·4 for the relative luminosity factor implies that radiation of wavelength 500 nm is 0·4 times as effective as radiation of wavelength 550 nm for the same power per unit area of radiation. Hence to give the same brightness the 500 nm radiation must transmit 1/0·4, *i.e.* 2·5 × (power/area) of the 550 nm radiation:

power per unit area $= 2\cdot5x$ watts per square metre

luminous flux $= (\text{power/area}) \times (\text{relative luminosity factor})$

$$= 2\cdot5x \times 0\cdot4$$

$$= x \text{ light-watts}$$

$$= 630x \text{ lumens}$$

This could of course have been obtained directly since the power/ area of the 500 nm radiation has been increased to give the same brightness and hence the same luminous flux as the 550 nm radiation.

Example 6.3

An incandescent lamp with a luminous flux of 3 900 lm is inserted in a sphere made up of a material which absorbs 10% of the flux. Assuming that the sphere distributes the flux uniformly in all directions what will be the luminous intensity of the sphere?

Luminous flux emitted by sphere is

$$\Phi = \frac{90}{100} \times 3\,900 = 3\,510 \text{ lm}$$

The solid angle containing this flux entirely surrounds the sphere, *i.e.* constitutes a larger sphere. Now the area of a sphere radius r is $4\pi r^2$. Hence the number of area sections each of area r^2 is 4π. One steradian is the solid angle which cuts off an area on the surface of a sphere, radius r, equal to r^2 units of area. A sphere thus constitutes 4π steradians. Hence luminous intensity will be $3\,510/4\pi$ lumens/ steradian = 279·3 lm/sterad = 279·3 cd.

Example 6.4

Calculate the luminance of the sphere in Example 6.3 if its radius is 12 cm and, assuming 15% of the luminous flux is incident on a table top of dimensions 1 m × 0·25 m, calculate the average illumination of the table top.

The apparent area of the spherical source in any direction equals that of a circle of radius 12 cm, *i.e.*

$$\text{Apparent area} = \pi \times (12 \times 10^{-2})^2 \text{ m}^2$$
$$= 144\pi \times 10^{-4} \text{ m}^2$$

The luminous intensity is 279·3 cd. Hence the luminance

$$= \frac{279\cdot3}{144\pi} \times 10^4 \text{ cd/m}^2$$

$$= 6\,172 \text{ cd/m}^2$$

The area of the table top = 0·25 m²

Incident luminous flux = 0·15 × 3 510 lm

Hence

$$\text{illumination} = \frac{0 \cdot 15 \times 3\,510}{0 \cdot 25} \, \text{lm/m}^2$$

$$= 2\,106 \, \text{lx}$$

PROBLEMS ON CHAPTER SIX

(1) Calculate the luminous flux in lumens of a light source emitting 10 W/m² of light radiation at wavelength 600 nm (assume the sensitivity at this wavelength is 0·5 and that one light watt is equivalent to 630 lm).

(2) Determine the frequency of and thus in what band and at which end of the band of the electromagnetic spectrum a radiation of wavelength 600 nm lies. What power per unit area is required of this radiation to give the same objective brightness as 5 W/m² of a radiation of wavelength 550 nm? Sensitivity factor of 600 nm radiation is 0·5.

(3) The amount of sun radiation incident upon a certain surface is 0·5 W/in², 40% of which is light radiation. It may be assumed that 1 W of this light radiation is equivalent to 0·3 light-watt. Calculate the illumination level of the surface in foot-candles.

(4) The average luminous intensity of a diffusing sphere surrounding an incandescent lamp having a total luminous flux (before absorbtion) of 5 000 lm is 300 cd. Determine the percentage of flux absorbed by the sphere.

(5) Calculate the luminance of the sphere in Problem 4 if its diameter is 35 cm and the average illumination of an area 0·5 m² if 20% of the luminous flux is incident upon it.

CHAPTER SEVEN

Circuit Quantities

7.1 CIRCUITS AND FIELDS

Electrical quantities will be discussed under two main headings: 'circuit quantities' and 'field quantities'. A circuit is defined, for the purposes of our discussion, as any collection of materials joined together in any way such that a common 'cause' sets up an 'effect' within the circuit as a whole. Three kinds of circuit will be discussed: the conductive circuit, in which cause and effect are taken as electromotive force and electric current respectively; the magnetic circuit, in which cause and effect are taken as magnetomotive force and magnetic flux respectively; and the electrostatic circuit, in which cause and effect are taken as electromotive force and electric flux respectively (*see* Fig. 7.1). All these quantities and further derived quantities will be discussed and clearly defined from first principles.

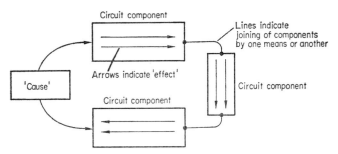

Fig. 7.1 Meaning of the term 'circuit'. (A simple series circuit is shown. It is applicable to all three types—Section 7.1.) Important quantities are 'cause' and 'effect' and their ratios.

A field is defined as the conditions existing within a *defined* part of a circuit, *i.e.* a part of the circuit whose physical dimensions (length and cross-sectional area) are taken into consideration in the quantities involved. Such quantities will be a measure of the distribution of the

82

'cause' over the length of the defined part, the density of the 'effect' throughout the area of the defined part and further quantities derived from these (*see* Fig. 7.2).

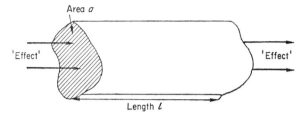

Fig. 7.2   Meaning of the term 'field'. A defined piece of the circuit of Fig. 7.1. Important quantities are 'cause per unit length' and 'effect per unit area' and their ratios.

7.2   THE ELECTRICALLY CONDUCTIVE CIRCUIT

An electrically conductive circuit is one through which electric charge is moved. The charge may be carried by electrons, ionised atoms or other particles, depending upon the nature of the material making up the circuit. This flow of electric charge is called an *electric current*. For the charge carriers to be freed from the material of which they are a constituent part and to move through the circuit they must be given energy. This energy may be acquired from any of several sources—chemical (batteries), heat (thermocouples), light (photovoltaic cells), electromechanical (generators), and so on.

7.3   ELECTROMOTIVE FORCE

The ability to cause a flow of electric current is measured in terms of the energy given to unit charge. Since the SI unit of energy is the *joule* and the unit of electric charge is the *coulomb*, the basic unit of the ability to cause a flow of electric current is the *joule/coulomb* (abbreviated J/C). It is important to realise that the number of charge *carriers* involved in carrying one unit of charge may or may not be unity depending upon the nature of the carrier. For example, if electrons are involved as the carriers a very great number of them are needed to carry one coulomb of charge ($6·3 \times 10^{18}$ in fact) since the charge per electron is so very small. The unit joule/coulomb is given a special name, the *volt* (abbreviated V), in honour of the Italian scientist Alessandros Volta, who did considerable original work on the nature of electrical energy sources. The term 'ability to cause a flow of electric current' is somewhat unwieldy and is in fact given the name *electromotive force*, abbreviated e.m.f. (symbol E).

Definition
Electromotive force is the ability to cause a flow of electric current and is measured in units of energy *given* to unit charge: joules per coulomb (J/C) or volts (V).

It will be appreciated that e.m.f. is *not* in fact a force since it is a measure of *energy* per unit charge.

7.4   ELECTRIC CURRENT

As was defined in Section 7.2, electric current is a movement or flow of electric charge. The unit of electric current is, quite logically, a measure of charge per unit time. The unit is the *coulomb/second* (C/s). One coulomb/second is called one *ampere* (abbreviated A) in honour of the French scientist who did research on a number of electrical and magnetic problems of his day.

The ampere is, in fact, the fourth fundamental unit of the SI, as was pointed out in Chapter 1, and the coulomb (which is the unit of the fourth fundamental concept) is defined using the unit of current, *i.e.* one coulomb of charge is the charge which is moved in one second when a current of one ampere flows; in other words, one coulomb is one ampere-second. How the fourth fundamental unit, the ampere, is defined, using magnetic principles, is described later in this chapter.

Definition
Electric current is a movement or flow of electric charge which is carried by some form of particle through the material through which the current is flowing. The unit of electric current is the coulomb/second (C/s) or ampere (A).

7.5   'CAUSE' AND 'EFFECT' IN THE CONDUCTIVE CIRCUIT

As has been indicated, the method of approach for all three types of circuit (conductive, magnetic and electrostatic) will be the same. In each case a 'cause' will be identified and the resulting 'effect' described. The cause in an electrically conductive circuit will be taken as e.m.f. and the effect as electric current.

7.6   OPPOSITION IN A CONDUCTIVE CIRCUIT

All circuits which will be examined have some form of inherent opposition to the setting up of the effect within the circuit. This opposition will depend upon, amongst other things, the nature and physical dimensions of the material or materials making up the circuit. To obtain some measure of the circuit opposition the cause is divided by the effect. By comparing the magnitude of the cause necessary to set up the same effect in circuits made of different materials a comparison of the quality of the material as a conductive, magnetic or electrostatic medium may be readily made.

In the conductive circuit the inherent circuit opposition is called *resistance* (symbol R) and a measure of the circuit resistance may be obtained by dividing the e.m.f. (cause) by the resultant current (effect):

$$\text{electrical resistance} \propto \frac{\text{e.m.f.}}{\text{current}}$$

As with other statements of proportionality, this statement can be made into an equation by the inclusion of a constant:

$$\text{resistance} = (\text{constant}) \times \frac{\text{e.m.f.}}{\text{current}}$$

and the constant may be defined to have any value.

In the SI the constant in such equations is taken as unity wherever possible in order to preserve, as far as one can, a coherent system. Taking the constant as unity,

$$\text{resistance} = \frac{\text{e.m.f.}}{\text{current}}$$

The unit of resistance is the *volt/ampere* (V/A). One volt/ampere is called one *ohm* in honour of the German scientist. The symbol for 'ohm' is the Greek letter omega, Ω.

Electrical resistance depends upon the material atomic structure and its physical dimensions and in certain special cases can vary with the applied e.m.f. This will be discussed in Chapter 8.

Definition
Electrical resistance is the opposition to current flow offered by a conductive circuit. The unit of resistance is the volt/ampere (V/A) or ohm (Ω).

7.7 SUPPORT IN A CONDUCTIVE CIRCUIT

As was stated in the preceding section, the opposition offered by a circuit made up of different materials may in turn be used as a guide to the quality of the materials as conductive, magnetic or electrostatic media. The inverse of opposition, *i.e. support*, offered by the circuit to the effect may also be used in a similar way. Clearly the greater the opposition of a particular circuit the smaller will be the support. A measure of circuit support may be obtained by dividing the effect in the circuit by the circuit cause. In this way, by comparing the magnitude of the effect produced by a cause of the same magnitude in circuits made of different materials, a comparison of the quality of the materials as conductive, magnetic or electrostatic media may be made. It is clear that since support is the inverse of opposition it will also depend on the nature of the material, its physical dimensions, and so on, as does the opposition although, naturally, the relationship between support and the material characteristics will be different from the relationship between opposition and the material characteristics.

The inherent circuit support of an electrically conductive circuit is called *conductance* (symbol G). Conductance is the reciprocal of resistance and this being so is determined directly by the equation

$$\text{conductance} = \frac{\text{current}}{\text{e.m.f.}}$$

The unit of conductance is the *ampere/volt* (A/V). One ampere/volt is called one *siemens*. The symbol for siemens is S.

Definition
Electrical conductance is the support of current flow offered by a conductive circuit. The unit of conductance is the ampere/volt (A/V) or siemens (S).

7.8 POTENTIAL DIFFERENCE

The concept of energy was discussed in Chapter 4. Broadly speaking, energy, which is the ability to do work, may be divided into various types and classes. Classes of energy include mechanical, electrical, heat, light, and so on. Types of energy include kinetic energy (energy by virtue of movement) and potential energy (energy by virtue of position), so that one could regard a moving electric charge, for example, as having electrical kinetic energy, and many other examples may be considered. The classification suggested is not rigid by any means but it does assist in forming a mental picture of a state of affairs not easily visualised.

Potential energy is energy by virtue of position relative to a fixed or datum level. If a body is held above ground it has potential energy and since no other effect is apparent it may be regarded as mechanical potential energy. The higher the body is held the greater is its potential energy with respect to ground level. Ground level is in this case the fixed or datum level. If the body is released the potential energy is converted to kinetic energy as the body is accelerated under the force of gravity and as its velocity increases so its kinetic energy (proportional to the square of the velocity) increases. The potential energy, which depends upon the distance between the body and the ground, is a measure of the energy capable of being converted to other forms. Potential energy of any class, therefore, can be regarded as a measure of convertible energy, its level depending upon the position of the body relative to a datum level.

The case so far discussed is one concerning mechanical potential energy but the idea can be extended to other forms. For example, if a bar is held with one end in a fire, the end in the fire is naturally hotter than the other end and a variation in temperature exists as one moves down the bar. Temperature was discussed in Chapter 5; it is a measure of the molecular heat energy of a body and here again a situation exists where the level of a measure of energy changes with position relative to a fixed level (in this case the end of the bar in the fire). It could be said that heat potential energy exists at different points along the bar.

Now consider the situation at present under examination, the electric circuit. Here a charge carrier has energy and the level and type of energy of the carrier with respect to a fixed level changes at different points around the circuit.

With reference to one end of a conductive circuit a charge carrier at the other end has maximum potential energy and since it is an electrical body the potential energy may be termed electrical potential energy. In the case of the conductive circuit the carrier actually moves and the electrical potential energy is converted to other forms— mainly to heat energy for a material offering reasonable opposition (solid conductors, etc.) or mainly to kinetic energy for a material offering minimal opposition (for example a cathode ray tube in which electrons are accelerated through a vacuum). Electrical potential energy then is a measure of the convertible energy of a charge carrier by virtue of its position relative to a fixed level. Whether this convertible energy is actually converted, as in the conductive circuit, or not, as in the electrostatic circuit (*see* Section 7.22), is irrelevant; the potential energy level still exists. The difference in levels of electrical potential energy at two points in a circuit whether conductive or electrostatic is called *potential difference*, abbreviated p.d. Since it is a measure of the difference of energy levels of a charge carrier it is measured in units of *energy* per unit *charge*, *i.e. joules* per *coulomb* or volts. This unit is, of course, the unit of e.m.f. The difference between the quantities e.m.f. and p.d. is only a subtle one; e.m.f. is a measure of the energy *given* to a charge carrier by an energy source (chemical, mechanical, heat, light, etc.), p.d. is a measure of the difference in potential energy levels of a charge carrier at two points in a circuit and the energy in this case is energy which is capable of being converted to other forms.

Definition
Potential difference between two points is a measure of the difference in the levels of the convertible energy of a charge carrier at the two points. It is measured in joules/coulomb (J/C) or volts (V).

Between the ends of a circuit the p.d. is at a maximum, *i.e.* it is the maximum convertible energy of the carrier. Since this is the energy

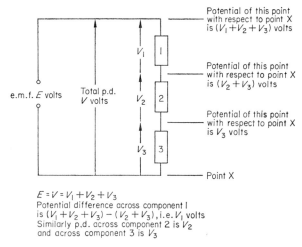

$E = V = V_1 + V_2 + V_3$
Potential difference across component 1
is $(V_1 + V_2 + V_3) - (V_2 + V_3)$, i.e. V_1 volts
Similarly p.d. across component 2 is V_2
and across component 3 is V_3

Fig. 7.3   E.m.f. and p.d.

made available by the source, *i.e.* the e.m.f., the p.d. between circuit ends is equal to the e.m.f. The sum of the p.d.'s taken across successive parts of the circuit is equal to the total p.d. or applied e.m.f. (*see* Fig. 7.3).

7.9   OHM'S LAW

As has been shown, the total resistance of a conductive circuit is determined by dividing the e.m.f. by the current. For a complete circuit e.m.f. is equal to the p.d. between the circuit ends or, as is usually written, the p.d. *across* the circuit, so it is equally true that the resistance of the complete circuit is equal to the total p.d. across the circuit divided by the total current flowing in the circuit.

As dividing the energy converted per unit charge in the *whole* circuit, *i.e.* total p.d., by the current gives the resistance of the *whole* circuit, so dividing the energy converted per unit charge in *part* of the circuit, *i.e.* the p.d. across the *part*, by the current flowing in the *part* gives the resistance of the *part*.

For any part of a conductive circuit the p.d. across the part V (volts), the current flowing in the part I (amperes) and the part resistance R (ohms) are related by the equation

$$R = \frac{V}{I}$$

This equation is an expression of *Ohm's Law*. It can also be written

$$V = IR \quad \text{and} \quad I = \frac{V}{R}$$

Since the conductance of a circuit or part of a circuit is equal to the reciprocal of resistance *i.e.* $G = 1/R$, where G is the conductance (siemens), then the three ways of writing Ohm's Law may be extended to:

$$G = \frac{I}{V} \quad \text{or} \quad V = \frac{I}{G} \quad \text{or} \quad I = VG$$

The appropriate form is chosen depending upon which quantities are known and which quantity it is required to determine.

Example 7.1

Determine the quantity not given in the following set of observations. V, I, R and G represent p.d., current, resistance and conductance respectively.

V (volts)	I (amperes)	R (ohms)	G (siemens)	
$V = IR$ or $V = I/G$	$I = V/R$ or $I = VG$	$R = V/I$ or $R = I/G$	$G = I/V$ or $G = I/R$	Solution
5	3	?	?	$R = \frac{5}{3}\,\Omega$ $G = \frac{3}{5}\,\text{S}$
4	?	2	?	$I = \frac{4}{2}\,\text{A}$ $G = \frac{1}{2}\,\text{S}$
?	8	6	?	$V = 8 \times 6\,\text{V}$ $G = \frac{1}{6}\,\text{S}$
?	3	?	5	$V = \frac{3}{5}\,\text{V}$ $R = \frac{1}{5}\,\Omega$
8	?	?	6	$I = 8 \times 6\,\text{A}$ $R = \frac{1}{6}\,\Omega$
7	2	?	?	$R = \frac{7}{2}\,\Omega$ $G = \frac{2}{7}\,\text{S}$
8	?	4	?	$I = \frac{8}{4}\,\text{A}$ $G = \frac{1}{4}\,\text{S}$

7.10 SUMMARY OF CONDUCTIVE QUANTITIES STUDIED SO FAR

TABLE 7.1

Quantities in the conductive circuit

Cause	Effect	Opposition (cause/effect)	Support (effect/cause)
e.m.f. E volts (V) or total p.d. V volts (V)	Current I amperes (A)	Resistance R ohms (Ω)	Conductance G siemens (S)

For a complete circuit or any part of a circuit the quantities p.d. V (volts), current I (amperes), resistance R (ohms) and conductance G (siemens) are related by Ohm's Law.

7.11 MAGNETISM

Having now completed a preliminary study of conductive circuit quantities the magnetic circuit will be similarly examined. Firstly it is necessary to discuss the phenomenon of magnetism and to define terms and special names used in the study of the subject.

Magnetism is defined as a property of certain materials whereby a force is set up between them and certain other materials. Whether the force is attractive or repulsive depends on the nature of the materials concerned and in what state of magnetism they are.

A magnetic material is one which by suitable treatment may be made to exhibit magnetism. The treatment necessary is given a special name magnetisation and a magnetic material thus treated is said to be magnetised.

A non-magnetic material is one which cannot be magnetised and does not experience a force when placed in proximity to a piece of magnetised material. The special name given to a piece of material which has been magnetised is magnet.

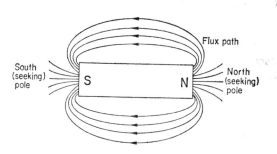

Fig. 7.4   Field pattern of a bar magnet.

The area of influence surrounding a magnet or other source of magnetism is called a magnetic field. The Earth behaves like a very large magnet with a magnetic field surrounding it. Small bar magnets react with the Earth's magnetic field and tend to set in a line closely approximating to one drawn between the north and south poles of the Earth. The end of a bar magnet which points to the north is called the north-seeking or north pole of the magnet; similarly the opposite end is called the south-seeking or south pole of the magnet.

Magnetic lines of force or flux lines are lines drawn to show the direction of action of a magnetic field, *i.e.* the line in which a small bar magnet would set if placed in the field. Lines of force are drawn from the north pole to the south pole of a magnet and the direction north to south indicated by an arrowhead (*see* Fig. 7.4). The number of lines of force or amount of flux can be used to indicate the relative strength of the magnetic field they are being used to diagrammatically represent.

7.12 ELECTROMAGNETISM

Magnetism and magnetic materials have been known and used since the time of the ancient Greeks. Early sailors used the interaction between the magnetic field of the Earth and that of small pieces of magnetic material (called lodestone) taken from caves to determine in which geographical direction they lay, although it is apparent they had no clear understanding of the theory behind their primitive navigational aid. For many years the two sciences of electricity and magnetism were treated separately although the English scientist Gilbert did a considerable service to their study as related subjects with his work in the latter half of the seventeenth century. However, it was not until 1823 when André-Marie Ampère, after whom the unit of electric current is named, produced his current ring theory of magnetism that the two sciences began to come together as one. Ampère's theory that the magnetism exhibited by certain materials was due to electric currents circulating within them has been shown by subsequent discoveries to be substantially correct. Michael Faraday indicated conclusively that there is a magnetic field associated with every electric current and atomic theory has shown that all materials contain moving electrons and, as has been shown, a movement of charge by a carrier constitutes an electric current. The difference between materials which behave magnetically and those which do not is explained by their molecular structure and the way in which the magnetic fields associated with the moving electrons either add to or cancel one another.

7.13 THE MAGNETIC FIELD OF A CONDUCTOR CARRYING CURRENT

A movement or flow of electric charge is called an electric current; any material which carries an electric current has a magnetic field associated with it. Use of a device such as a small bar magnet, *i.e.*

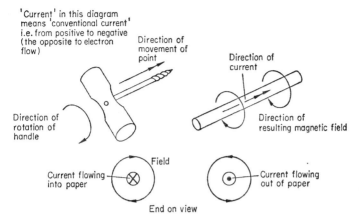

Fig. 7.5   Field surrounding a conductor—the corkscrew rule.

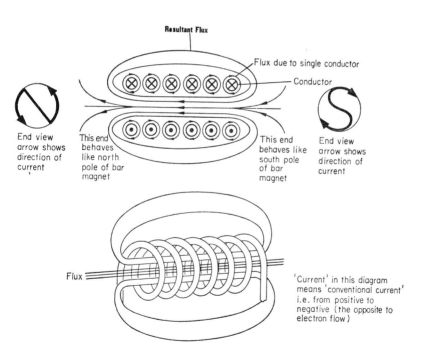

Fig. 7.6   Field due to a coil—compare Fig. 7.4.

a compass needle, or the sprinkling of very small particles of a magnetic material upon a piece of paper or card upon which the current-carrying material is lying shows that the lines of force of the field are in annular rings about the conductor. A useful device to aid recall of the direction of the field of a conductor carrying current is the so-called 'Corkscrew Rule', where the direction of rotation of the handle of a corkscrew to produce a movement of the point is the direction of the lines of force of the conductor field for a current flowing in the direction of movement of the point (*see* Fig. 7.5).

The magnetic field associated with a current-carrying conductor may be greatly increased by winding the conductor in the form of a coil. The lines of force of each turn add together to give a much increased flux through the centre of the coil and along its axis (*see* Fig. 7.6).

7.14   THE MAGNETIC CIRCUIT

A magnetic circuit is defined as any collection of materials joined together such that a common source of magnetism causes a magnetic field, and hence flux, to be set up within the confines of the materials making up the circuit. Figure 7.7 shows a typical form of magnetic

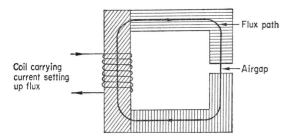

Fig. 7.7   A magnetic circuit.

circuit containing four kinds of material (three solids and air), in which the flux is set up within the solid material and the air gap by a coil of wire (through which a current is passed) wrapped round one of the limbs of the circuit. The flux is confined to the circuit because the materials shown (except for the air gap) are assumed to be magnetic materials. Air is not a magnetic material. This will be discussed more fully in succeeding sections.

7.15 MAGNETOMOTIVE FORCE

The magnetic lines of force or flux in a magnetic circuit such as that shown in Fig. 7.7 are set up by the current-carrying conductor wrapped in the form of a coil round one of the limbs of the circuit. The ability to set up a magnetic flux in such a case is related to the size of the electric current in the coil and to the number of coil turns, since varying either or both of these quantities varies the flux within the circuit. This ability is in fact measured in terms of these quantities in the SI. A query may well come to mind here as to how such a measurement applies when the magnetic field is set up by a device such as a magnet in which the source of the field, though electrical in origin (the intermolecular movement of electrons), is not easily measured in terms of current and coil turns. In this case measuring the ability to set up magnetic flux of a magnet in terms of current and coil turns is in fact comparing the ability of the magnet to that of a suitable coil and current to produce the same effect as the magnet.

The ability to set up magnetic flux is called *magnetomotive force* (abbreviated m.m.f.) and the unit is the *ampere-turn* (symbol At*). The symbol for m.m.f. is F. The m.m.f. of any coil of N turns carrying a current I amperes is IN ampere-turns.

To further clarify the point raised above concerning a permanent magnet, *i.e.* one not having a coil carrying an electric current associated with it, if it is stated that the m.m.f. of a permanent magnet is, for example, 1 000 At it is meant that the magnet m.m.f. is equivalent to that produced by a coil of N turns carrying a current I amperes such that $IN = 1\ 000$. Suitable figures would include a 250 turn coil carrying 4 A, a 1 000 turn coil carrying 1 A, a 500 turn coil carrying 2 A, and so on.

It is clear that m.m.f. is not in fact a force since its units are those of (current) × (turns), *i.e.* (coulomb/second) × (turns).

Definition

Magnetomotive force, m.m.f., symbol F, is the ability to set up a magnetic flux. The unit is the ampere-turn (At) and for a coil of N turns carrying a current I amperes, F is given by $F = IN$ ampere-turns.

---

* Since the turn is dimensionless, *i.e.* its physical size is unimportant and is thus not related to the basic units mass, length or time, it is often omitted from the symbol abbreviation which is then written A.

7.16 MAGNETIC FLUX

In preceding sections the term 'lines of flux' has been used synonymously with 'lines of force' and these have been defined as a diagrammatic means of representing the magnitude and direction of a
magnetic field. The quantity m.m.f. has been defined as the ability
to set up a magnetic flux. To talk of an ability to set up something
which is only a method of diagrammatic representation is inconsis-
tent with clear and precise thinking and it is therefore clear that
further discussion of the term 'magnetic flux' is required.

It is extremely difficult to formulate a precise definition of flux;
rather like the quantity electric charge it is one which is recognised
by its effects, is measured by its effects and in fact its unit is defined
by its effects. Magnetic flux may be loosely regarded as a state of
magnetic stress within or around a material, the word stress not
being applied in its precise mechanical sense, the greater the flux
the greater being the stress. It is important to recognise and appreciate
that flux is a state of being rather than a physical movement of
anything (for example, electric current is a movement of electrically
charged particles). The size of the unit of magnetic flux in the SI
can be visualised and described using the mechanical force set up
between magnetic fields, although it is not defined in this way. How
it is defined will be discussed in Chapter 9. The symbol for magnetic
flux is the Greek letter Φ (phi).

7.17 THE MAGNITUDE OF THE UNIT OF MAGNETIC
FLUX

Magnetism has been defined as a property of certain materials
whereby a force is set up between them and certain other materials.
If two bar magnets are laid side by side a physical force is exerted
between them owing to the interaction of their magnetic fields.
Whether the force is attractive or repulsive is determined by the
relative positions of the north and south poles of the magnets and
on the relative directions of the two fields (see Fig. 7.8). It can be
seen that if the two fields act in the same direction a force of repulsion
exists and for fields in opposite directions the force is one of attraction. This is true whether the fields are due to permanent magnets
or set up by coils carrying current.

If a wire carrying current is placed in a magnetic field due to any
cause the field due to the wire and the other field will react and the
wire will move (see Fig. 7.9). It can be shown that the magnitude of
the force acting on the wire depends upon the current in the wire,

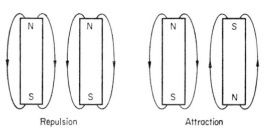

Repulsion   Attraction

Fig. 7.8   Force between magnets.

the length of the wire exposed to the magnetic field and the magnetic flux per unit area of the main field.

If the current is one ampere, the length of wire is one metre and the force is one newton, the main field has a flux of one unit per unit

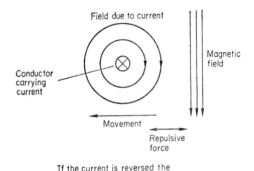

Field due to current

Conductor carrying current

Magnetic field

Movement

Repulsive force

If the current is reversed the direction of movement is reversed

Fig. 7.9   A wire carrying current in a magnetic field.

area. This is not how the unit is defined but it does give an idea of the size of the unit of flux.

The name of the unit of magnetic flux is the *weber* (abbreviated Wb) a special name again given in honour of a scientist.

7.18   DEFINITION OF THE AMPERE

The situation existing when two conductors carrying current exert a force upon one another owing to the interaction of their magnetic fields is used to define the unit of current the ampere.

If the force between two conductors carrying the same current, each one metre long and situated one metre apart, is 2×10^{-7} newtons, then the current in the conductors, by definition, is *one ampere.* The reason for the choice of the rather strange figure 2×10^{-7} is discussed in Appendix II.

7.19 'CAUSE' AND 'EFFECT' IN THE MAGNETIC CIRCUIT

The cause in a magnetic circuit will be taken as m.m.f., the effect as magnetic flux.

7.20 OPPOSITION IN THE MAGNETIC CIRCUIT

In the conductive circuit it was seen that the setting up of a flow of electric current is opposed to a certain extent, the magnitude of the opposition depending upon the physical dimensions and the nature of the material. Conductive circuit opposition or electrical resistance is measured for a complete circuit by dividing the 'cause' (the e.m.f.) by the 'effect' (the current).

Similarly, for magnetic circuits different materials react differently to an m.m.f. Some allow considerable flux, others a certain amount, whilst a great number offer so much opposition that the flux is negligible. As with the conductive case, magnetic circuit opposition depends upon the nature and physical dimensions of the material. It should be realised that what constitutes a good electrical conductor and supports electric current does not necessarily support magnetic flux in the same way. Copper, for example, is one of the best and most widely used conductors, but to all intents and purposes ranks as a non-magnetic material.

Magnetic circuit opposition is given a special name *reluctance* to distinguish it from the quantity resistance. The symbol is S. Reluctance is measured by dividing the m.m.f. (cause) by the flux (effect):

$$\text{reluctance} = \frac{\text{m.m.f.}}{\text{flux}}$$

and the unit is the *ampere-turn* per *weber* (At/Wb). This unit does not have a special name.

Definition

Magnetic reluctance is the opposition of a magnetic circuit to magnetic flux. The symbol is S and the unit is the ampere-turn/weber (At/Wb).

7.21 SUPPORT IN A MAGNETIC CIRCUIT

The support in a magnetic circuit is a measure of how well the circuit supports magnetic flux. It is the inverse of reluctance and is measured by dividing flux by m.m.f. (effect by cause). The special name for magnetic circuit support is *permeance*, the symbol for which is the Greek letter Λ (lambda). The unit of permeance is the *weber* per *ampere-turn* (Wb/At) for which there is no special name.

Definition
Magnetic permeance is the ability of a magnetic circuit to support magnetic flux. The unit is the weber/ampere-turn (Wb/At) and the symbol is Λ.

TABLE 7.2

Analogous quantities in conductive and magnetic circuits

Quantity / Circuit	Cause	Effect	Opposition (cause/effect)	Support (effect/cause)
Conductive	e.m.f. E volts (V) or total p.d. V volts (V)	Current I amperes (A)	Resistance R ohms (Ω)	Conductance G siemens (S)
Magnetic	m.m.f. F ampere-turns (At)	Flux Φ webers (Wb)	Reluctance S ampere-turns/ weber (At/Wb)	Permeance Λ webers/ampere-turn (Wb/At)

7.22 THE ELECTROSTATIC CIRCUIT

The electrostatic circuit is defined as a collection of materials joined together such that on the application of a common energy source no steady electric current flows after the initial movement of charge and a distribution of static charge then exists at points around the circuit (*see* Fig. 7.10).

To consider what does happen within the circuit, since no current flows electric charge will have to be discussed in more detail.

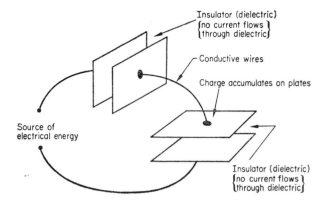

Fig. 7.10   An electrostatic circuit.

7.23   ELECTRIC CHARGE

As has been seen, electric charge is a difficult thing to define. Basically it must be regarded as a phenomenon of nature recognisable by its effects. These effects may be recognised by sight, sound, touch and even smell. The discharge of electricity during a storm can be seen and heard in the form of lightning and thunder; a discharge through the body with its associated effect on the body organs can be felt and the breakdown of certain gases leading to the formation of ozone can be sensed by the nose. The science of electric charge which is not moving, *i.e.* non-conductive, is called electrostatics.

The study of electrostatics began many thousands of years ago with the noticing of the phenomenon of charging by friction and by induction, and indeed the words electricity, electronics, electron, and so on have the common stem derived from 'electros', the Greek word for amber, a material particularly noticeable for its pronounced

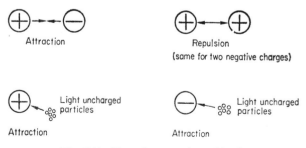

Fig. 7.11   Force between charged bodies.

electrostatic characteristics. It is now believed that when a body becomes charged it is either gaining or losing charge carriers, such as electrons, which once moved remain either on or off the body until the process is reversed, *i.e.* no continuous or steady-state electric current flows.

A body once charged exerts a force on another charged body, one of attraction if the two charges are opposite and of repulsion if the two charges are similar. Also a charged body exerts an attractive force on certain light uncharged materials (in the same way as a magnet attracts unmagnetised magnetic materials)—*see* Fig. 7.11.

7.24   THE ELECTRIC FIELD

The area surrounding an electrically charged body or between two or more charged bodies in which the influence of the charge can be felt is called an electric field. Charged bodies lying within the field will move if they are physically able to do so; if they do move the material thus forms part of a conductive circuit. It is clear that an electric field must exist within each material in a conductive circuit for the charge carriers to move. When an electric field exists within a material containing charge carriers and they are so physically bound that they cannot move the material is in a state of stress. If the electric field is strengthened continuously a point is reached at which physical breakdown occurs, the charge carriers break their bonds and an electric current flows. The breakdown point is determined by the nuclear forces binding the charge carriers to the material atoms.

As with the magnetic field, lines can be drawn to indicate the magnitude and direction of the field. Such lines are drawn from the positive end of the field to the negative end, *i.e.* the direction *opposite* to that in which an electron would move if released in the field (*see* Fig. 7.12). The lines indicating field magnitude and direction are called electric lines of force or electric flux lines. Electric flux can be regarded as a state of stress within a material (again, the word

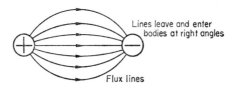

Fig. 7.12   The electric field.

stress not having its precise mechanical sense). The unit of electric flux will be discussed in a succeeding section.

7.25 THE p.d. ACROSS AN ELECTRIC FIELD

If a charge carrier is held within an electric field and released it will move, its direction depending upon the field direction (an electron, for example, would move towards a positively charged body). The force on the carrier depends upon its position in the field, the nature of the carrier and the size of its charge and the size of the charge on the charged body.

The energy the carrier would receive if it were released thus depends upon, amongst other things, its position in the field. This energy is therefore potential energy. The potential energy of a carrier varies at different places in the field and between any two points a difference in electrical potential energy levels may exist, *i.e.* a potential difference. The potential difference between any two points depends upon the field and thus the charge present in the area, the nature of the material in which the field exists and its physical dimensions. An electrostatic circuit is described as one in which no steady current flows and a distribution of static charge exists. Between the various levels of charge at different points round the circuit an electric field and a p.d. exist. As with the conductive circuit, the total p.d., *i.e.* the p.d. between ends of the circuit, since it is a measure of the total convertible energy per unit charge, is equal to the total available energy per unit charge, the e.m.f.

The sum of the p.d.s across successive points around the circuit is equal to the total p.d. and hence the e.m.f.

7.26 TRANSIENTS IN THE ELECTROSTATIC CIRCUIT

An electrostatic circuit consists of one or more non-conductive materials connected to an energy source (*see* Fig. 7.10). When the source is first applied the conditions existing are called transient conditions since they exist only for a short period of time.

When the source is first connected an electric field is set up within the circuit as a whole. Charge carriers capable of movement move in the field until they reach a point where they are no longer able to. At this point the carriers accumulate and the region becomes charged. The circuit is made up of one or more non-conductive materials joined together and to the energy source by conductive materials; thus current flows *only* within the conductive materials

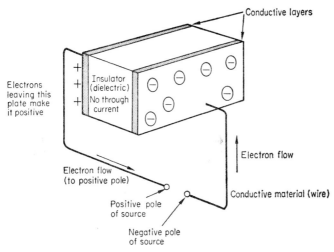

Fig. 7.13   Charging in an electrostatic circuit.

and charge accumulates in the end regions of the non-conductive materials (*see* Fig. 7.13). As the end regions become charged they repel further carriers (*e.g.* a region which has become charged by an accumulation of electrons will be negative and eventually repel further electrons) and hence movement of charge stops, *i.e.* the transient current dies away (*see* Fig. 7.14). The magnitude of the initial transient current is determined by the e.m.f. and the resistance of the conductive parts of the circuit.

Thus the e.m.f. connected to the circuit causes an initial or

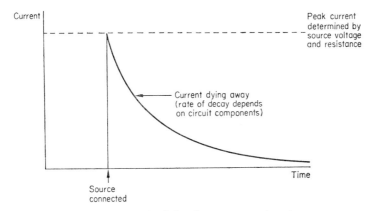

Fig. 7.14   Graph of charging current against time.

transient current to flow which in turn distributes charge about the circuit. Between the various charge levels, *i.e.* across the individual circuit materials, a p.d. exists and an electric field exists within them.

7.27 'CAUSE' AND 'EFFECT' IN THE ELECTROSTATIC CIRCUIT

The cause in an electrostatic circuit as a whole is of course e.m.f. since this sets up the initial electric field and charge distribution. The steady state or non-transient effect is the electric field and consequently electric flux.

Since the e.m.f. and total p.d. are equal the total p.d. may also be considered the cause in the circuit as a whole. For a part of the circuit the p.d. is taken as the cause and the flux within the part as the effect. This is logical since the p.d. is a measure of convertible energy per unit charge and is dependent in turn upon the whole circuit cause, the total available energy per unit charge, the e.m.f.

7.28   THE UNIT OF ELECTRIC FLUX

The unit of electric flux is taken as the flux due to a single unit of electric charge. It is convenient to give it the same name as the unit of charge, the coulomb. No confusion exists since with every unit of charge the appropriate level of electric flux exists. The symbol for electric flux is Ψ (psi).

7.29   OPPOSITION IN THE ELECTROSTATIC CIRCUIT

As with the other circuits considered, dividing cause by effect, *i.e.* e.m.f. or total p.d. by flux for a whole circuit or p.d. by flux for part of a circuit, gives a measure of the opposition of a material to the setting up of an electric flux within it. As with materials in the other circuits, a 'cause' of the same magnitude applied to different materials (of the same physical dimensions) sets up different levels of 'effect'. In this case, for example, the same e.m.f. applied to similar size pieces of paraffin wax, glass, copper, wood, etc. will set up different flux levels.

Paraffin wax and glass, for example, are good supporters of flux and thus offer little opposition. Copper, on the other hand, though it is a good conductor is not as good as paraffin wax (which is not a conductor) at supporting an electric flux. The difference in the conductive characteristic lies in the fact that only a weak electric field is necessary in copper for its charge carriers to be freed and to move. The charge carriers in paraffin wax or glass (both *insulators*) are very tightly bound to the parent atoms and a very strong field is necessary before conductive breakdown occurs.

Opposition to flux in a non-conductive material may be measured in units of e.m.f. or p.d. per unit flux, *i.e.* volts per coulomb (V/C). The unit has no special name nor does the characteristic of opposition since it is not generally considered in engineering as a useful quantity.

7.30 SUPPORT IN THE ELECTROSTATIC CIRCUIT

Dividing effect by cause gives a measure of support. In the electro-static case the unit of support is the *coulomb* per *volt* (C/V). One coulomb per volt is called one *farad* after Michael Faraday, the 'father' of electrical engineering. The characteristic of support of electric flux is called *capacitance* and a device specially made to have high support is called a *capacitor* (formerly a condenser but this name is not in use any more)—*see* Fig. 7.15. The capacitance of a circuit as a whole or of part of a circuit is obtained by dividing the

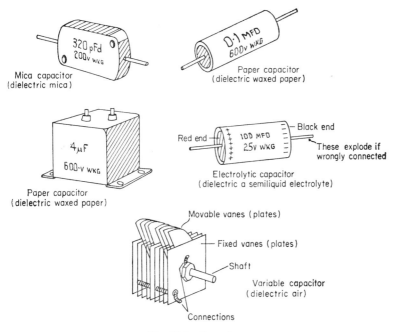

Fig. 7.15 Capacitors.

flux by the e.m.f. or total p.d. (for the whole circuit) or p.d. (for a part). It is a measure of the ability of the circuit or circuit part to support flux.

Since electric flux is associated with the corresponding level of electric charge, capacitance may also be regarded as the ability of a circuit or circuit part to become charged and store charge and a capacitor may be considered as a device specially built to store electric charge.

Definition
Capacitance is the ability of a material to support electric flux or to store charge. It is measured in coulombs per volt (C/V) or farads (symbol F).

TABLE 7.3

Analogous quantities in various circuits

Quantity / Circuit	Cause	Effect	Opposition (cause/effect)	Support (effect/cause)
Conductive	e.m.f. E volts (V) or total p.d.	Current I amperes (A)	Resistance R ohms (Ω)	Conductance G siemens (S)
Electrostatic	Potential difference V volts (V)	Flux Ψ coulombs (C) (charge indicated by Q)		Capacitance $C = \Psi/V$ or Q/V coulombs/volt (farad, F)
Magnetic	Magneto-motive force F ampere-turns (At)	Flux Φ webers (Wb)	Reluctance S ampere-turns/ weber (At/Wb)	Permeance Λ webers/ampere-turn (Wb/At)

7.31 SOME WORKED EXAMPLES ON CHAPTER 7

Example 7.2
The p.d. across a certain material is 6 J/C. If the current flowing is 5 C/s find the rate of energy dissipation in joules/second.

$$\text{Rate of energy dissipation} = 6 \text{ J/C} \times 5 \text{ C/s}$$
$$= 30 \text{ J/s}$$

Note that this is of course power (*see* Chapter 4) measured in watts (one watt is one joule/second), so that power in watts (J/s) = p.d. in volts (J/C) × current in amperes (C/s).

Example 7.3

Two materials A and B are formed into pieces of the same length and cross-sectional area. A p.d. of 10 V is applied across each in turn and currents of 10 A and 5 A flow through A and B respectively. Which material has the greater conductance?

$$\text{Conductance of A} = \frac{\text{current}}{\text{p.d.}} = \frac{10}{10} = 1 \text{ S}$$

$$\text{Conductance of B} = \frac{\text{current}}{\text{p.d.}} = \frac{5}{10} = 0{\cdot}5 \text{ S}$$

A has the greater conductance.

Example 7.4

Five pieces of conductive material of resistances 10 Ω, 15 Ω, 25 Ω, 30 Ω and 20 Ω respectively are connected end to end successively to form a complete circuit (this is called a series connection). The current flowing through the circuit (through each material in turn) is 5 A when a certain e.m.f. is applied. What is the value of the e.m.f.?

The p.d. across each part by Ohm's Law is given by (current) × (resistance). Hence the p.d.s across the 10 Ω, 15 Ω, 25 Ω, 30 Ω and 20 Ω parts are 5 × 10 V, 5 × 15 V, 5 × 25 V, 5 × 30 V and 5 × 20 V respectively, *i.e.* 50 V, 75 V, 125 V, 150 V and 100 V. Hence the total p.d. which equals the sum of the p.d.s across the circuit parts is equal to (50 + 75 + 125 + 150 + 100) V, *i.e.* 500 V. Therefore, since e.m.f. = total p.d., e.m.f. = 500 V.

Example 7.5

Calculate the reluctance of a circuit in which the magnetic flux is 600 mWb and the m.m.f. is supplied by a coil of 500 turns carrying 3 A.

$$\text{m.m.f.} = (\text{turns}) \times (\text{current})$$
$$= 500 \times 3$$
$$= 1\ 500 \text{ At}$$
$$\text{flux} = \frac{600}{1\ 000} \text{ Wb}$$

Hence

$$\text{reluctance} = \frac{\text{m.m.f.}}{\text{flux}}$$
$$= \frac{1\,500 \times 1\,000}{600}$$
$$= 2\,500 \text{ At/Wb}$$

Example 7.6
An e.m.f. of 150 V is applied to a coil of 1 000 turns of total resistance 300 Ω. Calculate the magnetic flux in the coil centre if the permeance of the coil core is 0·001 Wb/At.

$$\text{Current in coil} = \frac{\text{e.m.f.}}{\text{resistance}}$$
$$= \frac{150}{300} \text{ A}$$
$$= 0·5 \text{ A}$$
$$\text{m.m.f.} = (\text{turns}) \times (\text{current})$$
$$= 1\,000 \times 0·5$$
$$= 500 \text{ At}$$

Since

$$\text{permeance} = \frac{\text{flux}}{\text{m.m.f.}}$$
$$\text{flux} = (\text{permeance}) \times (\text{m.m.f.})$$
$$= 0·001 \times 500 \text{ Wb}$$
$$= 0·005 \text{ Wb}$$
$$= \frac{5}{1\,000} \text{ Wb}$$
$$= 5 \text{ mWb}$$

Example 7.7
Calculate the electric flux between the plates of a 0·1 μF capacitor across which there is a p.d. of 200 V.

The capacitance is 0·1 μF, i.e. 0·1 × 10⁻⁶ F, i.e. 0·1 × 10⁻⁶ C/V, and the p.d. is 200 V. Therefore

$$\text{flux} = 0·1 \times 10^{-6} \times 200 \text{ C}$$
$$= 20 \text{ μC}$$

Since 1 C of flux exists due to 1 C of charge this is also the charge on the capacitor.

Example 7.8

Find the capacitance of a capacitor if a p.d. of 250 V sets up a flux (or charge on the plates) of 1 mC.

$$p.d. = 250 \text{ V}$$

$$\text{charge (or flux)} = 10^{-3} \text{ C}$$

Therefore

$$\text{capacitance} = \frac{10^{-3}}{250} \text{ F}$$

$$= \frac{1\,000}{250} \times 10^{-6} \text{ F}$$

$$= 4 \,\mu\text{F}$$

PROBLEMS ON CHAPTER SEVEN

(1) A current of 5 C/s flows through a certain piece of material across the ends of which there is a p.d. of 20 J/C. Calculate (a) the energy converted in one minute, (b) the conductance of the material, (c) the resistance of the material, (d) the current (amperes) if the p.d. is trebled.

(2) The energy converted in 25 s in a piece of material A by a p.d. of 15 V is 2 500 J and in a piece of material B by a p.d. of 20 V is 3 000 J. Calculate the conductance of each material and determine which material is the better resistor.

(3) The p.d. across a 1 500 turn coil of resistance 20 Ω is 100 V. Calculate the permeance of the coil core if the resultant flux is 20 mWb.

(4) A 2 000 turn coil has a flux of 200 mWb in its core when 10 A flows. Calculate the reluctance of this core and the core of Problem 3 and determine which is the better magnetic material.

(5) Calculate the charge on a 10 μF capacitor if the p.d. is the same as that which when applied to a resistor of 200 Ω causes a current of 10 A to flow.

CHAPTER EIGHT

Field Quantities

8.1 GENERAL DISCUSSION

A field has been defined (Section 7.1) as the conditions existing within a defined part of a circuit, *i.e.* a part of the circuit whose length and cross-sectional area are taken into consideration in the quantities involved. Two kinds of field have been discussed: the electric field through which charge carriers either move if the material is conductive or stay in a state of stress if the material is non-conductive, and the magnetic field which sets up a 'magnetic stress' within the material in the field.

Quantities so far discussed are circuit quantities only, *i.e.* quantities which in general describe the conditions of a circuit or part of a circuit without reference to the physical dimensions of the materials involved in the circuit or circuit part. The quantities termed field quantities take into consideration the physical dimensions of the material. As will be seen in succeeding sections and examples, not taking physical dimensions into consideration can lead to erroneous first impressions of the conductive or resistive characteristics of a material.

8.2 GENERAL TREATMENT

For all three fields—the electrically conductive field, the electrically non-conductive field and the magnetic field—the treatment will be the same. The quantities obtained by dividing the previously discussed 'cause' by field length, 'effect' by field cross-sectional area and the ratio of these quantities one to the other will be examined and the implications of the new quantities thus obtained will be considered. These new quantities produced by ratio will be related to circuit quantities already discussed.

8.3   THE ELECTRIC FIELD: VOLTAGE GRADIENT AND ELECTRIC FIELD STRENGTH

The electric field involved in both conductive and non-conductive materials is, of course, the same field. The difference in the two cases is the 'effect', the one being a movement of charge, termed electric current, the other being a state of stress, termed electric flux.

The first field quantity examined is common to both conductive and non-conductive cases; it is termed voltage gradient or *electric field strength* and, as the name implies, is a measure of the strength of the electric field.

As indicated in Section 8.1, it is not sufficient to use the p.d. across the ends of a material lying in an electric field to indicate the strength of the field within the material. On surface examination 10 000 V indicates a stronger field than 1 000 V but this is not in fact so, if, for example, the 10 000 V exists across 1 cm of a material and the 1 000 V exists across 1 mm of a material—the two fields are the same in magnitude. The length over which the field is acting must be taken into consideration.

Electric field strength, symbol E, is determined by dividing p.d. by the length across which the p.d. exists (or e.m.f. by the length of complete circuit for the electric field strength in a complete circuit). The unit is therefore the *volt/metre* (V/m):

$$\text{electric field strength} = \frac{\text{e.m.f. or p.d.}}{\text{length}} \text{ V/m} \qquad (8.1)$$

8.4   ELECTRIC CURRENT DENSITY

As the size of the p.d. is insufficient to judge electric field strength, so electric current alone is not sufficient if one is interested in the distribution of the effect in various materials. Cross-sectional area must be taken into account. *Electric current density* is defined as current divided by area and the unit is the *ampere/square metre* (A/m^2) or, more commonly, ampere/square millimetre (A/mm^2):

$$\text{current density} = \frac{\text{current}}{\text{area}} \text{ A/m}^2 \qquad (8.2)$$

Electric current density is of interest mainly to power engineers who are concerned with the transmission of power at a relatively high current. The symbol is J.

Electric field strength and current density are illustrated diagrammatically in Fig. 8.1.

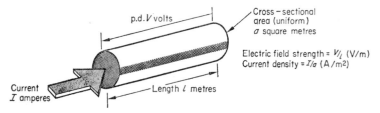

Fig. 8.1   Illustrating current density and electric field strength (Section 8.5).

8.5   ELECTRICAL RESISTIVITY

After 'cause' and 'effect' had been examined in the study of circuit quantities the quantities 'opposition' and 'support' were discussed for each circuit type. Similarly here, now cause and effect in a defined part of a circuit have been studied giving cause/unit length and effect/area, the ratios of these quantities will be examined giving opposition and support for a defined part or *specific* opposition and support.

The ratio of cause per unit length to effect per unit area gives a measure of the opposition for a defined part or specific opposition.

For the conductive case, electric field strength (V/m) divided by current density (A/m^2) gives a quantity measured in

$$\frac{\text{volts}}{\text{amperes}} \times \frac{(\text{metre})^2}{\text{metre}}$$

or ohm-metres. Clearly this is a measure of resistance but with a physical dimension included. To appreciate the significance of the dimensions consider a piece of conductive material of length l metres, cross-sectional area a square metres, carrying a current I amperes and with a p.d. of V volts across the length (*see* Fig. 8.1). For this piece of material

$$\text{electric field strength} = \frac{V}{l} \text{ (volts/metre)}$$

$$\text{current density} = \frac{I}{a} \text{ (amperes/metre}^2\text{)}$$

and the ratio

$$\frac{\text{electric field strength}}{\text{current density}} = \frac{V}{I} \times \frac{a}{l}$$

but V/I is the resistance of the material through the length l. Denoting this by R (ohms),

$$\frac{\text{electric field strength}}{\text{current density}} = R \times \frac{a}{l}$$

If a is one square metre and l one metre, *i.e.* a *cube* of side *unity* (*unit cube*), of material, the RHS of this equation is equal to the resistance between opposite faces.

The magnitude of this quantity depends on the material. It is called the *resistivity* or *specific resistance* of the material and is denoted by ρ (rho). Hence

$$\text{resistivity} = \frac{\text{electric field strength}}{\text{current density}} \tag{8.3}$$

and

$$\rho = R \times \frac{a}{l}$$

Hence

$$R = \frac{\rho l}{a} \tag{8.4}$$

This equation, which is an important one, states the dependence of the resistance of any piece of material of any length or cross-sectional area on three things:

resistance $\propto$ resistivity, *i.e.* the type of material

resistance $\propto$ length

resistance $\propto \dfrac{1}{\text{area}}$

TABLE 8.1

Resistivity (specific resistance) in microhm millimetres
(resistance between opposite faces of a metre cube)

Silver	16·6	
Copper	17·8	
Gold	24·2	
Aluminium	32·1	
Tungsten	50	Conductors
Zinc	61	
Brass	66	
Platinum	110	
Silicon	6×10^5	Semi-conductor
Glass	5×10^{18}	
Porcelain	2×10^{22}	Insulators
Paraffin wax	3×10^{25}	

This in turn implies that the greater the length of a material the greater is the electrical resistance and the greater the cross-sectional area the smaller is the resistance. This seems a logical state of affairs and is accepted without further discussion.

The resistivity of a material can be taken as a measure of its effectiveness as an insulator or inversely as a conductor. Some examples are given above. It should be noted that the unit of resistivity is the ohm-metre (Ω m) since in the dimensions given by the equation, (ohm-metre)^2/metre, the dimension of the denominator may be cancelled by part of the dimensions of the numerator. The preferred unit uses the sub-multiples microhm millimetre, abbreviated $\mu\Omega$ mm.

The resistivity of a material is substantially constant at constant temperature and for most materials is independent of voltage and current.

8.6   ELECTRICAL CONDUCTIVITY

Dividing 'cause per unit length' by 'effect per unit area' gives a measure of the *opposition per unit cube*. In the conductive case this means electric field strength divided by current density gives resistivity. Dividing 'effect per unit area' by 'cause per unit length' gives the reciprocal of the above, the *support per unit cube*. In the conductive case this means current density divided by electric field strength gives a quantity called *conductivity* or *specific conductance*, denoted by the symbol σ (sigma). As can be inferred, conductivity is the conductance between opposite faces of a metre cube.

Considering the same conditions as in the previous section, *i.e.* a piece of material of length l metres, cross-sectional area a square metres, carrying a current I amperes and with a p.d. across it of V volts,

$$\text{electric field strength} = \frac{V}{l} \text{ (volts/metre)}$$

$$\text{current density} = \frac{I}{a} \text{ (amperes/metre}^2)$$

Hence

$$\text{conductivity } (\sigma) = \frac{\text{current density}}{\text{electric field strength}} \qquad (8.5)$$

$$= \frac{I}{V} \times \frac{l}{a}$$

$$= G \times \frac{l}{a} \text{ siemens/metre}$$

where G is the conductance of the material in siemens. Hence

$$G = \frac{\sigma a}{l} \qquad (8.6)$$

and, as before, if a is one square metre and l is one metre, $G = \sigma$, where G is the conductance between opposite faces of a metre cube. Hence the conductivity or specific conductance is the conductance between opposite faces of a metre cube.

Typical values of conductivity may be determined by calculating the reciprocals of the figures for resistivity given in Table 8.1. Note that the unit of conductivity will be the siemens/metre. Again, the preferred unit in general use is the kilosiemens per millimetre (kS/mm).

TABLE 8.2

Conductive field quantities

Cause/length	Effect/area	Opposition per unit cube	Support per unit cube
Electric field strength E (V/m) (e.m.f. or p.d./length)	Current density J (A/m^2) (current/area)	Resistivity ρ (Ω m)	Conductivity σ (S/m)

8.7 SOME WORKED EXAMPLES ON THE CONDUCTIVE FIELD

Example 8.1
A piece of material A has a cross-sectional area of 0·1 cm^2, length 50 cm and carries of current of 3 A. A piece of material B has a cross-sectional area of 0·5 cm^2, length 15 cm and carries a current of 5 A. The p.d. across each material is 10 V. Determine which material is the better conductor.

It is insufficient merely to compare the conductance or resistance of the two pieces, as such a comparison merely determines which piece has the greater conductance (or smaller resistance). It does not determine which material is the better conductor. To do this the relative physical dimensions must be taken into consideration along with the respective p.d. and current levels.

To demonstrate this, consider the conductance of each piece in turn. Conductance is determined by dividing current by p.d. Hence conductance of the piece of material A is 3/10, *i.e.* 0·3 S, the conductance of the piece of material B is 5/10, *i.e.* 0·5 S. The piece of material B thus has the greater conductance.

Now consider the conductivity of each material. As shown above, conductivity is given by

$$\frac{(\text{conductance}) \times (\text{length})}{\text{area}}$$

The length and area must be in metres and square metres respectively. For material A

$$\text{conductivity} = \frac{0.3 \times 50 \times 10^{-2}}{0.1 \times 10^{-4}}$$

$$= 15\,000 \text{ S/m}$$

$$= 15 \times 10^{-3} \text{ kS/mm}$$

For material B

$$\text{conductivity} = \frac{0.5 \times 15 \times 10^{-2}}{0.5 \times 10^{-4}}$$

$$= 1\,500 \text{ S/m}$$

$$= 1.5 \times 10^{-3} \text{ kS/mm}$$

As can be seen, material A has the greater conductivity and is thus the better conductor. Logically it might be expected that this is so since, although the piece of material A is three times greater in length and five times smaller in cross-sectional area than the piece of material B, it is carrying 3/5 of the current that the piece of material B carries, *i.e.* the current is not reduced proportionally.

Example 8.2
A piece of material resistance 20 Ω is stretched to four times its original length, whilst the original volume is maintained. What is the new value of resistance?

The new length is four times the original length and, since resistance is proportional to length, it will be increased by a factor of four. Since the volume remains the same and volume is equal to (length) × (cross-sectional area), if the length is increased by a factor of four, the area is reduced by a factor of four. Resistance is inversely proportional to area, hence the resistance will be increased by a factor of four. Thus the new resistance will be 20 × 4 (due to increased length) × 4 (due to reduced area), *i.e.* 320 Ω.

This solution assumes a uniform material which is stretched uniformly so that the area is proportionally reduced throughout the length.

Example 8.3

Would either or both of the materials listed below break down if submitted to a p.d. of ten million volts across one metre length?

Air (breakdown field strength 3 000 V/mm)
Porcelain (breakdown field strength 15 000 V/mm)

Ten million volts per metre is an electric field strength of

$$\frac{10 \times 10^6}{1} \text{ V/m}$$

The breakdown field strength of air is 3 000 V/mm, *i.e.* 3×10^6 V/m and of porcelain is 15 000 V/mm, *i.e.* 15×10^6 V/m.

Hence the air would break down but the porcelain would not. This example indicates one of the reasons why porcelain-based materials are used in the support of power lines. Voltages of the order of one million are not, of course, transmitted along the lines but voltages of this order and above are encountered should lightning strike the line. It is then essential to prevent breakdown between line and pylon.

8.8 MAGNETIC FIELD QUANTITIES, MAGNETIC FIELD STRENGTH

The same methods of analysis will be employed for the magnetic field as for those with the conductive field. It will be recalled that 'cause' and 'effect' in a magnetic circuit are m.m.f., measured in ampere-turns, and magnetic flux, measured in webers, respectively.

Magnetic field strength is determined by dividing m.m.f. (the cause) by the length of the material in which the field is set up. The symbol is H and the unit is, of course, *ampere-turn* per *metre* (At/m):

$$\text{magnetic field strength} = \frac{\text{m.m.f.}}{\text{length}} \text{ At/m} \qquad (8.7)$$

8.9 MAGNETIC FLUX DENSITY

The distribution of the 'effect', *i.e.* magnetic flux, within a material is called magnetic flux density. It is determined by dividing the flux in webers by the cross-sectional area in square metres. The symbol is B

and the unit *webers* per *square metre* (Wb/m²). One weber per square metre is called one tesla, symbol T.

$$\text{magnetic flux density} = \frac{\text{magnetic flux}}{\text{area}} \text{ (Wb/m}^2) \text{ or (T)} \quad (8.8)$$

See Fig. 8.2 for an illustration of *B* and *H* in the magnetic field.

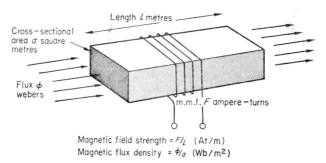

Fig. 8.2   Illustrating flux density and magnetic field strength (Sections 8.9 and 8.8).

8.10   MAGNETIC OPPOSITION PER UNIT CUBE

Dividing magnetic field strength by magnetic flux density gives a measure of the opposition of a material to the setting up of a magnetic flux between the opposite faces of a unit cube. The quantity would be analogous to electrical resistivity in the conductance field. It is not, however, used by engineers and consequently will not be discussed further.

8.11   MAGNETIC SUPPORT PER UNIT CUBE, PERMEABILITY

Dividing magnetic flux density by magnetic field strength gives a measure of the support of a material to the setting up of a magnetic flux between the opposite faces of a unit cube. The quantity is analogous to electrical conductivity in the conductive field and is of prime importance in the study and choice of suitable magnetic materials for motors, generators, transformers and other electromagnetic devices. It is called magnetic *permeability* and is symbolised by the Greek letter μ (mu):

$$\text{permeability} = \frac{\text{magnetic flux density}}{\text{magnetic field strength}} \quad (8.9)$$

The unit is obtained by an examination of the units of the quantities from which permeability is derived. Magnetic flux density has the unit Wb/m^2, magnetic field strength the unit At/m; hence the unit of permeability is

$$\frac{Wb}{m^2} \times \frac{m}{At} \quad \text{or} \quad \frac{Wb}{At} \times \frac{1}{m}$$

i.e. weber per ampere-turn metre. A more convenient way of expressing this is discussed in Chapter 9 (Section 9.5).

Expressing eqn. (8.9) symbolically

$$\mu = \frac{B}{H}$$

$$= \frac{\Phi}{a} \times \frac{l}{F}$$

where Φ is flux (Wb), F is m.m.f. (At), l is length (m) and a is area (m^2)

$$= \frac{\Phi}{F} \times \frac{l}{a}$$

i.e.

$$\mu = \Lambda \times \frac{l}{a}$$

where Λ is permeance (*see* Section 7.22). Hence

$$\Lambda = \mu \times \frac{a}{l} \tag{8.10}$$

This equation is directly comparable with eqn. (8.6). In general it may be expressed verbally as (circuit support) = (support per unit cube) × (area/length).

Permeability of a material is usually expressed as a product of two numbers, the permeability of a vacuum (*absolute permeability* μ_0) and the ratio of the permeability of the material to the permeability of a vacuum (*relative permeability* μ_r). Thus

$$\mu = \mu_r \times \mu_0$$

The value of μ_0 is fixed in the International System as $4\pi \times 10^{-7}$ units.* The choice of values for μ_0 in different unit systems is discussed briefly in Appendix II; μ_r may vary between values about unity for non-magnetic materials to values as high as 100 000 for

---

* The units are actually H/m as explained in the next chapter and μ_0 is usually given as $0.4\pi \ \mu H/m$.

magnetic materials. Broadly speaking, materials are classified magnetically by their values of relative permeability as follows.

Diamagnetic materials have a value of μ_r less than unity and become very weakly magnetised in a direction opposite to the applied field. This classification includes bismuth, antimony, copper, zinc, mercury, gold and silver.

Paramagnetic materials have values of μ_r slightly above unity and become weakly magnetised in the direction of the applied field. This classification includes aluminium, platinum, oxygen, air, manganese and chromium. For the purposes of the discussion in this book both paramagnetic and diamagnetic materials are considered non-magnetic.

Ferromagnetic materials become strongly magnetised in the direction of the applied field and have values of μ_r ranging from a few hundreds up to a hundred thousand. Some of these materials are annealed iron (pure), μ_r from 200 to 5 000, silicon iron, μ_r from 600 to 10 000 and permalloy (78 % nickel, 22 % iron), μ_r up to 100 000.

Unlike its analogous conductive field quantity conductivity, the permeability of magnetic materials depends to some considerable degree upon the magnetic flux density and magnetic field strength (conductivity for *most* conductive materials is substantially constant at constant temperature and is independent of voltage or current levels).

8.12   THE VARIATION OF PERMEABILITY, B/H CURVES

In reading that magnetic flux density in a material for a particular magnetic field strength depends upon permeability which in turn depends on the flux density one is reminded of the chicken and egg problem (which came first?).

The reason for this apparent anomaly is saturation. As the magnetic field in a material is progressively increased the rate of increase of flux becomes steadily less owing to the material becoming saturated. It is believed that a magnetic material may be considered to be made up of a considerable number of 'molecular magnets' which line up with the applied field. Once the total number is completely aligned the material is saturated and a further increase in flux by increasing the applied field becomes impossible (*see* Fig. 8.3). For a material of fixed length and area it follows that the ratio of magnetic flux density to magnetic field strength becomes progressively smaller as saturation is reached. A typical graph plotting B against H is shown in Fig. 8.4 where this point is illustrated. Before a choice of magnetic material for any particular purpose is made B/H curves for the

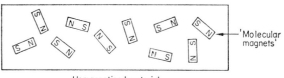

Unmagnetised material

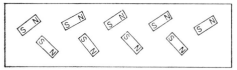

Partially magnetised material

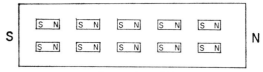

S | N

Saturated material

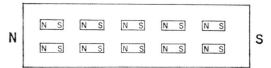

N | S

Saturated material (reverse of above)

Fig. 8.3   'Molecular magnets'.

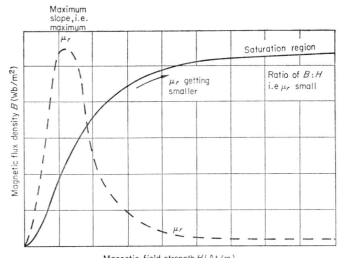

Fig. 8.4   Graph of B/H and μ_r/H.

material are consulted and the magnetic characteristics are considered bearing in mind the purpose to which the material is to be put.

TABLE 8.3

Analogous field quantities

Quantity / Circuit	Cause/length	Effect/area	Opposition per unit cube	Support per unit cube
Conductive	Electric field strength E (V/m) (e.m.f. or p.d./ length)	Electric current density (A/m²) (current/area)	Electric resistivity ρ (Ω m)	Electric Conductivity σ (S/m)
Magnetic	Magnetic field strength H (At/m) (m.m.f./length)	Magnetic flux density B (Wb/m²) (tesla, T) (magnetic flux/ area)	Not usually considered	Permeability μ (Wb/At m) ($\mu = \mu_r\mu_0$) ($\mu = B/H$)

Worked examples on these quantities will be found in Section 8.17.

8.13   THE NON-CONDUCTIVE OR STATIC ELECTRIC FIELD

The final field to be considered is essentially the same as the conductive field considered earlier with the important difference that no charge carriers move within the field and, consequently, though the 'cause' is the same, *i.e.* e.m.f. or total p.d. for a complete circuit or p.d. across the part for part of a circuit, the 'effect' is different and will be electric flux instead of electric current.

The 'cause per unit length' is the same as for the conductive field, has the same name and symbol and is determined in the same way, *i.e.*

$$\text{electric field strength, } E = \frac{\text{e.m.f. or p.d.}}{\text{length}} \text{ (V/m)}$$

8.14   ELECTRIC FLUX DENSITY

'Effect per unit area' for the static electric field is electric flux density and is determined by dividing electric flux (in coulombs) by area (in square metres); the symbol is D and the unit *coulombs* per *square metre* (C/m^2):

$$\text{electric flux density} = \frac{\text{electric flux}}{\text{area}} \, (C/m^2) \qquad (8.11)$$

Electric flux density is of course analogous to magnetic flux density (Wb/m^2) for the magnetic field and current density (A/m^2) for the conductive electric field (*see* Fig. 8.5).

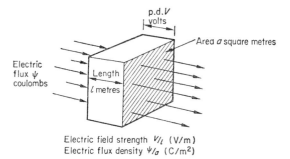

Fig. 8.5 Illustrating electric field strength and electric flux density (Sections 8.13 and 8.14).

8.15 OPPOSITION TO ELECTRIC FLUX PER UNIT CUBE

Dividing 'cause/length' by 'effect/area' gives a measure of the opposition to the effect between opposite faces of a unit cube (a cube of side one metre). In this case the opposition is to the setting up of electric flux in a cube and would be determined by dividing electric field strength by electric flux density. In practice, as with the analogous quantity in the magnetic field, this quantity is not used.

It should be noted that the only quantity of this type which is used in the three fields is resistivity (*see* Section 8.5), which is the opposition to the setting up of an electric current between opposite faces of a cube of side one metre.

8.16 SUPPORT FOR ELECTRIC FLUX PER UNIT CUBE, PERMITTIVITY

Dividing electric flux density (effect per unit area), by electric field strength (cause per unit length) gives the support of a unit cube to the setting up of electric flux between opposite faces. This quantity is called electric *permittivity* and its symbol is ε (epsilon):

$$\text{permittivity} = \frac{\text{electric flux density}}{\text{electric field strength}} \qquad (8.12)$$

Writing this symbolically,

$$\varepsilon = \frac{D}{E}$$

i.e.

$$\varepsilon = \frac{\Psi}{a} \times \frac{l}{V}$$

where Ψ is electric flux (coulombs), V is p.d. (volts), l is length (metres) and a is area (square metres). Hence

$$\varepsilon = \frac{\Psi}{V} \times \frac{l}{a}$$

The quantity Ψ/V, *i.e.* flux per unit p.d., is measured in coulombs per volt or *farads* and is called capacitance, symbol C. Hence the unit of permittivity is farad-metre/(metre)^2 or farad/metre (F/m). Hence

$$\varepsilon = C \times \frac{l}{a} \text{ (farads/metre)}$$

and therefore

$$C = \frac{\varepsilon a}{l} \qquad (8.13)$$

Equation (8.13) is directly comparable with eqn. (8.6) for the conductive case and eqn. (8.10) for the magnetic case (*see* Section 8.11).

As with (magnetic) permeability the (electric) permittivity ε of a material is written as the product of two numbers, one being the permittivity of a vacuum (*absolute permittivity* ε_0), the other being the ratio of the permittivity of the material to the permittivity of a vacuum (*relative permittivity* ε_r). Thus

$$\varepsilon = \varepsilon_0 \times \varepsilon_r$$

The value of ε_0 is fixed in the International System as 8·85 pF/m. The choice of this value is discussed, together with the choice of value of absolute permeability μ_0, in Appendix II.

Relative permittivity ε_r varies, depending upon the material, from approximately unity (air) to values in the thousands (certain ceramic materials). Some examples are glass (ε_r between 4 and 10), paper (ε_r between 2 and 3 depending on treatment), porcelain (ε_r between

4 and 7), tantalum oxide (ε_r about 28), titanium dioxide (ε_r up to 170) and ceramics (ε_r anywhere between 5 and 4 000).

It can be seen from eqn. (8.13) that capacitance for fixed length and area of insulator is directly proportional to permittivity and hence to relative permittivity. Relative permittivity of a medium is therefore also a ratio of the capacitance between two plates with the medium as a separator to the capacitance of the same plates with a vacuum (or, approximately, air) occupying a space of the same length and cross-sectional area as the separator. The insulator in a capacitor is called a *dielectric*, hence ε_r is often called the *dielectric constant*.

TABLE 8.4

Analogous quantities in fields

Circuit \\ Quantity	Cause/length	Effect/area	Opposition per unit cube	Support per unit cube
Conductive	Electric field strength E (V/m) (e.m.f. or p.d./length)	Electric current density (A/m^2) (current/area)	Electric resistivity ρ (Ω m)	Electric conductivity σ (S/m)
Magnetic	Magnetic field strength H (At/m) (m.m.f./length)	Magnetic flux density B (Wb/m^2) (tesla, T) (magnetic flux/area)	Not usually considered	Permeability μ (Wb/At m) ($\mu = \mu_r\mu_0$) ($\mu = B/H$)
Electrostatic	Electric field strength E (V/m) (e.m.f. or p.d./length)	Electric flux density D (C/m^2) (electric flux/area)	Not usually considered	Permittivity ε (F/m) ($\varepsilon = \varepsilon_r\varepsilon_0$) ($\varepsilon = D/E$)

8.17 SOME WORKED EXAMPLES ON FIELD QUANTITIES

Other worked examples on conductive field quantities only are in Section 8.7.

Example 8.4

Calculate the current flowing in a coil of 1 000 turns if the core has a flux density of 0·03 Wb/m^2 and the relative permeability at this level is 10 000. The core length is 20 cm.

Permeability $\mu = \mu_r \times \mu_0$. Hence for the core

Also
$$\mu = 10\,000 \times 4\pi \times 10^{-7} \text{ Wb/At m}$$

$$\mu = \frac{B}{H}$$

where B is the flux density (Wb/m^2), H is the magnetic field strength (At/m). Hence

$$H = \frac{B}{\mu}$$

In this case

$$H = \frac{0\cdot03}{10\,000 \times 4\pi \times 10^{-7}} \text{ At/m}$$

Now

$$\text{magnetic field strength} = \frac{\text{m.m.f.}}{\text{length}}$$

Therefore

$$\text{m.m.f.} = H \times (\text{length})$$

$$= \frac{0\cdot03}{10\,000 \times 4\pi \times 10^{-7}} \times 0\cdot2 \text{ At}$$

Therefore

$$\text{current flowing} = \frac{\text{ampere-turns}}{\text{turns}}$$

$$= \frac{0\cdot03 \times 0\cdot2}{10\,000 \times 4\pi \times 10^{-7}} \times 1\,000 \text{ A}$$

$$= 0\cdot48 \text{ mA}$$

It can be seen that with such a relatively small flux density, a coil of many turns and a core of such high permeability only a small current is required.

Example 8.5
Calculate the reluctance of a core of length 50 cm, cross-sectional area $0\cdot01 \text{ cm}^2$ and relative permeability 5 000. What m.m.f. would be required to set up a flux of $0\cdot01$ mWb in this core?

From eqn. (8.10)

$$\text{permeance} = \frac{(\text{permeability}) \times (\text{area})}{\text{length}}$$

Reluctance (circuit opposition) is the reciprocal of permeance (circuit support). Hence

$$\text{reluctance } (S) = \frac{\text{length}}{(\text{permeability}) \times (\text{area})}$$

Therefore

$$S = \frac{0\cdot5}{5\,000 \times 4\pi \times 10^{-7} \times 0\cdot01 \times 10^{-4}} \text{ At/Wb}$$

$$= 79\cdot5 \times 10^6 \text{ At/Wb}$$

Note that length must be expressed in metres, area in square metres. $79\cdot5 \times 10^6$ At are required to set up a flux of 1 Wb in the core; hence 795 At would be needed to set up a flux of $0\cdot01$ mWb assuming the same value of relative permeability. This m.m.f. could be achieved with $0\cdot795$ A through 1 000 turns, $7\cdot95$ A through 100 turns, and so on.

Example 8.6
A parallel plate capacitor (*see* Fig. 7.15) has a dielectric $0\cdot1$ cm thick and cross-sectional area 10 cm^2 and a capacitance of $0\cdot001$ μF. What is the value of the dielectric constant? What would be the new capacitance if this insulator were replaced by one of the same size but with relative permittivity 8?

From eqn. (8.13)

$$C = \frac{\varepsilon a}{l}$$

where ε is permittivity, a is area (m^2), l is length (m). Hence

$$\varepsilon = \frac{Cl}{a}$$

and since $\varepsilon = \varepsilon_r \times \varepsilon_0$, where ε_r is the dielectric constant (relative permittivity),

$$\varepsilon_r = \frac{\varepsilon}{\varepsilon_0} = \frac{Cl}{\varepsilon_0 a}$$

$$= \frac{0\cdot001 \times 10^{-6} \times 0\cdot1 \times 10^{-2}}{8\cdot85 \times 10^{-12} \times 10 \times 10^{-4}}$$

$$= 11\cdot3$$

(since ε_r is a ratio it has no units).
    If the dielectric were replaced by one having the same physical dimensions but of relative permittivity 8, the capacitance would be

reduced in the ratio 11·3 to 8, *i.e.* if C_x is the new capacitance in microfarads

$$\frac{0·001}{C_x} - \frac{11·3}{8}$$

Therefore

$$C_x = \frac{8 \times 0·001}{11·3}$$

$$C_x = 0·0007 \, \mu F$$

Example 8.7

Calculate the charge stored by a capacitor having a dielectric of length 0·1 cm, area 50 cm² and dielectric constant 10 when the p.d. is 100 V.

To calculate charge it is first necessary to calculate capacitance:

$$\text{capacitance} = \frac{\varepsilon \times (\text{area})}{\text{length}}$$

$$= \frac{10 \times 8·85 \times 10^{-12} \times 0·1 \times 10^{-4}}{50 \times 10^{-4}} \, F$$

$$= 0·177 \times 10^{-12} \, F$$

$$= 0·177 \, pF$$

Capacitance is measured in farads; one farad is the special name given to the coulomb/volt where the coulomb, in this case, is the unit of flux. Since by definition one unit of flux is set up by one unit of charge, one coulomb of flux is set up by one coulomb of charge. Hence the coulomb in the unit of capacitance may also be taken as the charge stored by the capacitor.

The capacitance is $0·177 \times 10^{-12}$ F, *i.e.* $0·177 \times 10^{-12}$ coulombs/ volt, so for a p.d. of 100 V the charge stored is

$$0·177 \times 10^{-12} \times 100 \text{ coulombs}$$

i.e. 177×10^{-12} coulombs (or about 1 100 million electrons, purely as a matter of interest).

PROBLEMS ON CHAPTER EIGHT

(Take $\mu_0 = 0·4\pi \, \mu H/m$, $\varepsilon_0 = 8·85 \text{ pF/m}$.)

(1) Calculate the current flowing in a coil of 2 500 turns if the core has a flux density of 0·4 T and the relative permeability at this level is 15 000. The core length is 25 cm.

(2) Calculate the capacitance of a parallel plate capacitor having a plate area 15 cm^2 and plate separation 0·01 cm if the dielectric constant is 3. What is the value of the electric charge on the plates if a p.d. of 500 V is applied?

(3) A piece of material of area 0·01 m^2 and length 0·5 m and having a value of specific conductance of 0·1 A/V is subjected to a p.d. of 150 V. Calculate the energy converted within the material in 100 s.

(4) A piece of material of resistance 150 Ω is stretched to four times its original length whilst maintaining constant volume. The resultant wire is then wound in the form of a coil having 5 000 turns. Calculate the m.m.f. of this coil if a p.d. of 200 V is applied.

(5) Two capacitors A and B are connected in series (so that the charge on each is the same) across a supply of 225 V d.c. The p.d. across capacitor B is 150 V and the capacitance of capacitor A is 16 μF. Calculate the capacitance of capacitor B.

CHAPTER NINE

Electromagnetic Induction

9.1 INTRODUCTION AND HISTORY

To select a single name from the many scientists who contributed to the knowledge of electromagnetism during the years of discovery from 1800 (when Oersted first wrote an account of his experiments concerning the connection between electricity and magnetism) onwards through the 19th century, would be difficult and, in fact, unfair to the memory of those who were omitted. However, for the topic under discussion in this chapter it is reasonable to recall the names of four of the greatest names in early electrical engineering. These are Oersted, the observer and originator, Ampère, the analyser, whose current ring theory of magnetism was discussed briefly in Section 7.12, Arago, whose now celebrated disc experiment (*see* Fig. 9.1) laid the ground for further investigation into why an apparently non-magnetic material such as copper should, when moving, affect separate magnetic materials nearby and which, in

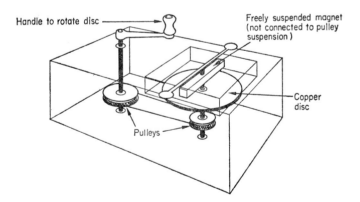

Handle to rotate disc

Freely suspended magnet
(not connected to pulley
suspension)

Copper
disc

Pulleys

Fig. 9.1 Form of Arago's disc experiment. When the copper disc is rotated the magnet, which is not mechanically coupled in any way to the disc, also rotates.

turn, led to the discovery and formulation of the laws of electro-magnetic induction by the English scientist, Michael Faraday.

In 1831, Faraday constructed his now famous ring apparatus, an iron ring on which were wound a number of coils of different numbers of turns. He imagined that it might be possible to convert magnetism into electricity from his knowledge of Arago's disc experiments and considerable further original thought. He was puzzled by the inter-mittent current which flowed in the unconnected winding when he applied a source of energy to another winding on the same core. It is not clear from his records what exactly he did anticipate but one assumes he thought the current should be steady once it had been established. He continued his experiments for some time and finally realised that the intermittent nature of the current was due to the changing magnetic field in the core linking the two coils and the field changed only when connecting or disconnecting the energy source. His discovery than an e.m.f. can be *induced* across a con-ductor by a changing field firstly explained Arago's disc—the moving copper disc was cutting the Earth's magnetic field setting up small circulating currents (eddy currents) in the disc and these currents in turn produced a local magnetic field which reacted with the magnetic materials nearby—and eventually led to modern engineering achieve-ments such as generators, certain types of motor, transducers and transformers.

9.2 FARADAY'S LAW OF ELECTROMAGNETIC INDUCTION

An e.m.f. is induced across a conductor situated in a changing magnetic field. The conductor may move through a field of constant magnitude fixed in space as in d.c. and small a.c. generators; a field of constant or varying magnitude may move past a conductor (or set of conductors) fixed in space as in the alternator and certain other electrical machines, or the conductor or set of conductors and field may be fixed in space but the magnitude of the field may vary as in the transformer. In all cases the condition that a conductor is situated in a changing magnetic field is satisfied. (In the case where the magnitude of the field is fixed but mechanical movement occurs, *i.e.* either field or conductors move, the field is changing with respect to the conductor by having its lines of force cut by the conductor.)

In all cases the lines of force of the field should either be at right angles or have a component at right angles to the end-to-end direction of the conductors.

For a coil of N turns the induced e.m.f. is proportional to N and to the rate of change of magnetic flux. Written symbolically, taking e as induced e.m.f. (in volts), t as time (in seconds), Φ as magnetic flux (in webers),

$$e = N \left(\frac{d\Phi}{dt} \right) \text{ volts} \qquad (9.1)$$

($d\Phi/dt$ is the mathematical way of writing 'rate of change of flux with respect to time').

A scientist named Lenz, who conducted further experiments on the nature of the induced e.m.f., came to the conclusion that the e.m.f. would act in a direction so as to *oppose* the change causing it. This is illustrated in Fig. 9.2. A conductor is being moved into a field whose

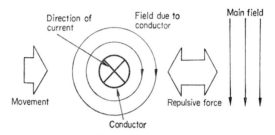

Fig. 9.2   Illustrating Lenz' law (Section 9.2). So as to set up repulsive force, current flows into the paper causing a field in the same direction as the main field (*see* also Fig. 7.8).

lines of force act vertically. Should the conductor form part of a closed circuit a current will flow setting up its own magnetic field. This field will be in a direction such that a force of repulsion exists between the main field and that of the conductor (for further clarification examine Fig. 7.8 and read Section 7.17 again). This force due to the induced e.m.f. opposes the motion causing the e.m.f. and must be overcome in continuing the motion. Owing to the nature of opposition of the induced e.m.f. it is called a *back e.m.f.* and is symbolised by a negative sign. Thus eqn. (9.1) is more correctly written as

$$e = -N \left(\frac{d\Phi}{dt} \right) \qquad (9.2)$$

This is Faraday's Law.

Example 9.1
Calculate the e.m.f. induced across a 500 turn coil cutting a magnetic field of flux 0·05 Wb in 0·5 s.

Average rate of flux

$$= \frac{0 \cdot 05}{0 \cdot 5} \text{ Wb/s}$$

$$= 0 \cdot 1 \text{ Wb/s}$$

Hence induced e.m.f.

$$= -500 \times 0 \cdot 1 \text{ V}$$

$$= -50 \text{ V}$$

It should be noted that in fact $d\Phi/dt$ means the instantaneous rate of change of flux and the solution is only correct if the speed of movement of the coil is constant.

9.3   SELF INDUCTANCE

The phenomenon of electromagnetic induction is used to good effect in a.c. and d.c. generators, induction motors, transformers and a.c. 'choking coils', but it must be realised that it occurs, whether it is required or not, in all cases where the magnetic field surrounding a conductor is changing. As has been shown (Section 7.8 *et seq.*), there is always a magnetic field surrounding a conductor which is carrying electric current and the magnitude of the resultant flux depends amongst other things upon the magnitude of the current. This flux may, of course, be increased by making the surrounding (or enclosed in the case of a coil) medium of high permeability magnetic material and by winding the conductor in the form of a coil with many turns (*see* Fig. 7.6), but a flux exists whether or not these arrangements are made, although it is, of course, quite weak in the case of a straight conductor in air (a non-magnetic medium).

Whenever the current is changing, either by closing or opening a d.c. circuit or all the time for a closed a.c. circuit, the associated magnetic field is changing. Since this satisfies the conditions of electromagnetic induction (a changing field in a plane at right angles to the end-to-end direction of the conductor), a back e.m.f. is induced across the conductor in a manner so as to oppose the change causing it. In the case of d.c. circuits being opened or closed the back e.m.f. endeavours to maintain the current or prevent the current respectively. This can result in a spark occurring at the switch on opening, or a delay in the rise of current on closing, a d.c. circuit (*see* Fig. 9.3), the magnitude of both effects of the back e.m.f. being dependent upon its size and hence on the rate of change of the field causing it.

With a.c. circuits the current is constantly changing both in direction and magnitude and consequently a back e.m.f. exists all the time the current flows. This back e.m.f. opposes the source e.m.f. to some extent and consequently cuts down the maximum value of a.c. reached in each direction in the circuit. This explains why a coil of wire with a magnetic core allows a far greater current on d.c. (once established) than a.c. for the same effective e.m.f.; this principle is used in 'choking coils' or chokes.

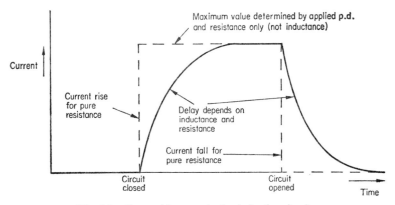

Fig. 9.3   Current/time graphs for inductive circuit.

The phenomenon of a conductor setting up its own induced e.m.f. is called *self inductance*, symbol L, and it is defined in the following manner. Since the back e.m.f. depends upon the rate of change of flux it also depends on the rate of change of current, *i.e.* the back e.m.f.

$$e \propto \frac{\mathrm{d}i}{\mathrm{d}t} \tag{9.3}$$

where i represents current in amperes, t represents time in seconds and, as before, $\mathrm{d}i/\mathrm{d}t$ represents the rate of change of current with time.

The *coefficient of self inductance* L is defined as the quantity which converts the statement of proportionality (9.3) to the equation

$$e = -L \left(\frac{\mathrm{d}i}{\mathrm{d}t}\right) \tag{9.4}$$

[The negative sign, as in eqn. (9.2), indicates the opposing nature of the induced e.m.f., *i.e.* that it is a back e.m.f.]

The unit of self inductance, the *Henry* (symbol H), is defined using eqn. (9.4): if the current in a conductor (or set of conductors) is changing at the rate of *one ampere* per *second* and the induced e.m.f. across the ends of the conductor (or set of conductors) is *one volt* then by definition the self inductance of the conductor (or set of conductors) is *one henry.*

Example 9.2

The current in a 10 H choke is raised from zero to 5A in 0·5 s. Calculate the average induced e.m.f.

Average rate of change of current per unit time is given by 5/0·5 A/s, *i.e.* 10 A/s. Hence

$$\text{induced e.m.f.} = -L \left(\frac{\mathrm{d}i}{\mathrm{d}t}\right)$$

$$= -10 \times 10$$

$$= -100 \text{ V}$$

Again the nature of this value should be noted, *i.e.* that it is a mean value since a mean rate of change has been taken. In other words, a linear current change has been assumed. In general di/dt means the *instantaneous* rate of change of current with time (*see* Example 9.1). In practice, of course, the current does *not* rise linearly (*see* Fig. 9.3).

Definition

The quantity *self inductance*, symbol L, is the ability of a conductor or set of conductors to induce an e.m.f. across itself when the current through the conductor(s) changes. It is measured in *henrys*, symbol H (note the plural of henry).

9.4   THE DIMENSIONS OF THE UNIT OF INDUCTANCE

From eqns. (9.2) and (9.4)

$$e = -N \left(\frac{\mathrm{d}\Phi}{\mathrm{d}t}\right) \quad \text{and} \quad e = -L \left(\frac{\mathrm{d}i}{\mathrm{d}t}\right)$$

hence

$$N \left(\frac{\mathrm{d}\Phi}{\mathrm{d}t}\right) = L \left(\frac{\mathrm{d}i}{\mathrm{d}t}\right)$$

and

$$L = N \left(\frac{\mathrm{d}\Phi}{\mathrm{d}i}\right) \qquad (9.5)$$

or, in words,

self inductance = (number of turns) × (rate of change of flux
with current)

The units of the right-hand side are *weber-turns* per *ampere*. Hence one *henry* means one *weber-turn* per *ampere*.

9.5   INDUCTANCE AND PERMEANCE COMPARED

Re-examine Section 7.21 concerning the quantity 'Support in a Magnetic Circuit'—*permeance*. The unit of permeance is the *weber* per *ampere-turn*.

'Turns' are dimensionless, *i.e.* they cannot be expressed in terms of the four basic quantities mass, length, time and charge as *all* other quantities (except temperature and luminous intensity) can be. So *dimensionally* the units of both inductance and permeance are the same—the weber/ampere. How this is written in terms of the four basic quantities of electromechanical engineering is discussed in Chapter 10. From this it appears that inductance and permeance (though not numerically equal) are the same quantity or at least indicative of the same state of affairs. Whether or not this is so will be discussed shortly.

Numerically, inductance, measured in weber-turns/ampere, would need to be divided by $(turns)^2$ to give permeance, measured in webers/ampere-turn, *i.e.*

$$\frac{\text{inductance}}{(\text{turns})^2} = \text{permeance} \qquad (9.6)$$

Now let us consider whether or not these quantities are in fact the same.

Permeance is a measure of the support of a magnetic circuit or part of a circuit offered to magnetic flux. In other words, the ease with which it is possible to set up a flux in the magnetic material forming the circuit or circuit part. Inductance, on the other hand, is a measure of the ability of a conductor (or set of conductors) to induce an e.m.f. across its ends when the current through it changes. On the face of it there seems little to connect the quantities until a more detailed examination of the process of self-induction is carried out. As has been shown, the e.m.f. is induced by the changing current but only because the changing current has set up a changing flux. The induced e.m.f. depends directly upon the magnitude of the rate of change of flux with current [eqn. (9.5)] and hence to some extent on the amount of flux produced per unit of current.

Since permeance is a measure of the flux produced per unit of current, self inductance and permeance are related and are both a measure of the same thing. So that as one regards permeance as a measure of the 'support' of a magnetic circuit or part of a circuit offered to magnetic flux, so one can equally regard inductance (which is a measure of the effect of this support) as a measure of the same thing. Hence conductance, inductance and capacitance may be regarded as analogous quantities in that they represent the support of the conductive, magnetic and electrostatic circuit (or circuit part) respectively.

As discussed above, the units of inductance and permeance are dimensionally the same, for, since the quantity 'turns' cannot be expressed in terms of the fundamental quantities, it is dimensionless (*see* Chapter 10) and thus the weber-turn/ampere (unit of inductance) and the weber/ampere-turn (unit of permeance) reduce to the weber/ampere, dimensionally the *henry*. Hence permeance may also be expressed in henrys and it follows that *permeability*, the support per unit cube (Section 8.11), having the unit weber per ampere-turn per metre, may be expressed in henrys/metre. This emphasises the analogy of the three measures of circuit support per unit cube, conductivity measured in siemens/metre, permeability in henrys/metre and permittivity in farads/metre, *i.e.* all three being expressed in units of circuit support/unit length.

9.6   THE EFFECT OF THE COIL CORE ON SELF INDUCTANCE

Equation (9.5) states that the self-inductance of a coil is equal to the number of turns on the coil multiplied by the rate of change of flux with current. Figure 8.4 shows a graph of the magnetic flux density B in a material plotted against the magnetic field strength H. Since both magnetic flux density and magnetic field strength are related to flux and magnetomotive force respectively by constants for a particular specimen (being cross-sectional area and length respectively) it follows that the flux/m.m.f. graph for a coil with a particular core material will have a similar shape to the B/H curve for the core. A typical flux/m.m.f. curve is shown in Fig. 9.4.

The quantity 'rate of change of flux with current' for any value of current is the *slope* or *gradient* of this curve at that value of current. Examination of the curve of Fig. 9.4 at three points, x, y, and z shows a fairly steep gradient, a shallower gradient and zero gradient respectively. Hence the inductance of a coil with a core of the material whose characteristics are displayed by this curve will depend upon

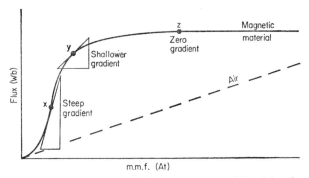

Fig. 9.4   Flux/m.m.f. curves for magnetic material and for air.

what standing value of current exists in the coil. On a.c., for example, where the current alternates about zero, the inductance changes all the time but its mean value will be the value corresponding to zero standing current, *i.e.* $N \times$ (slope of the curve at the origin). On d.c. the inductance again changes depending upon the value of the current at any time, its mean value depending on the mean value of the d.c. This discussion is sufficient to show that a coil with a magnetic core does not in fact have constant inductance. Figure 9.4 also shows a typical flux/m.m.f. graph for an air-cored coil and as one might expect since the phenomenon of saturation (Section 8.12) does not enter into it, the graph is in fact a straight line. The gradient of a straight line is constant, hence the rate of change of flux with current, and thus the inductance of an air-cored coil, is constant. However, since the core is virtually non-magnetic the flux achieved

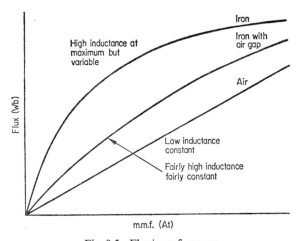

Fig. 9.5   Flux/m.m.f. curves.

per unit m.m.f. is very much reduced and hence the inductance is very much reduced.

In summary, an air-cored coil has a low constant inductance, an iron-cored coil has a high varying inductance. A suitable compromise to obtain a high constant inductance is effected by introducing an air gap into the iron core (*see* Fig. 9.5).

9.7 MUTUAL INDUCTANCE

As has been shown, an e.m.f. is induced across a conductor or set of conductors whenever the magnetic flux linking the conductors is changing. So far, in studying the phenomenon of self inductance the variation of magnetic flux has been due to the changing current in the conductors themselves. If two coils are suitably arranged in close proximity, such that flux due to the current in one coil links the other coil, changing the current in the first or *primary* coil changes the flux linking the two coils and an e.m.f. is induced across the second or *secondary* coil due directly to the changing current in the primary coil. The magnitude of this induced e.m.f. depends upon the rate of change of current in the primary coil, the strength of flux produced by the primary current and the respective numbers of turns on the primary and secondary coils. This phenomenon is called *mutual inductance*.

An equation relating the secondary e.m.f. with the rate of change or primary current is

$$e_2 = M \left(\frac{di_1}{dt} \right) \tag{9.7}$$

where the suffixes 1 and 2 indicate 'primary' and 'secondary' respectively and where M is the *coefficient of mutual inductance*. M is determined, as indicated above, by the turns ratio of the coils and the flux linkages produced by the changing current.

Compare eqn. (9.7) with eqn. (9.4). If a current changing at the rate of one *ampere* per *second* in the primary coil produces an induced e.m.f. across the secondary coil of one *volt* the mutual inductance M is said to have the value *unity* and from eqn. (9.4) it is clear that the unit is the *henry*.

As the coefficient of self inductance of a coil depends upon physical features such as turns, core material, etc., so also will the coefficient of mutual inductance of two magnetically coupled coils depend upon the turns *ratio*, the material linking the coils and its magnetic properties. The exact relationships will not be further discussed

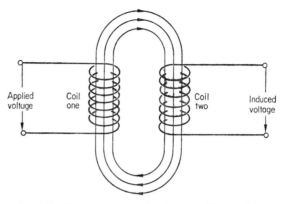

Fig. 9.6 Illustrating mutual inductance (Section 9.7).

since this text is primarily on fundamental quantities and their units (*see* Fig. 9.6).

The phenomenon of mutual inductance, which in fact was part of Faraday's original discoveries (Section 9.1), is used particularly in transformers.

9.8 SOME WORKED EXAMPLES ON CHAPTER 9

Example 9.3
Calculate the coefficient of self inductance of a coil if the rate of change of current producing an e.m.f. of 10 V is 0·5 A/s.

Since

$$\text{induced e.m.f.} = L \times (\text{rate of change of current})$$

$$10 = L \times 0·5$$

$$L = 20 \text{ H}$$

Example 9.4
Calculate the average value of the coefficient of self inductance of a 1 000 turn coil if the average value of e.m.f. induced across its ends is 2 V over a period of 5 s in which the current changes uniformly from 4 A to 6 A.

The average value of the rate of change of current is 2 A per 5 s, *i.e.* 0·4 A/s. Since the average e.m.f. is 2 V, the average value of *L*, the coefficient of self inductance, is given by

$$L = \frac{2}{0\cdot4} = 5\,\text{H}$$

Example 9.5
Using the information given in Example 9.4, determine the flux produced in the coil core by a current of 5 A flowing in the coil. Assume the core is being worked under the conditions of Example 9.4.

The assumption which is allowed means that the core is being worked on the same part of the flux/current curve and hence the calculated average value of the coefficient of self inductance may be used. If the core were being worked on any other part of the curve the value of L (and hence the induced e.m.f.) might change depending on the gradient of the flux/current curve (*see* Section 9.6 for clarification).
   The quantity required here is the connection between flux and m.m.f. or circuit support quantity *permeance*:

$$\text{permeance} = \frac{L}{(\text{turns})^2}$$

hence

$$\Lambda = \frac{5}{10^6}$$

($L = 5$ H and there are 1 000 turns on the coil)

$$= 5 \times 10^{-6}\,\text{Wb/At}$$

When 5 A flows the m.m.f. is 5 000 At and since

$$\text{permeance} = \frac{\text{flux}}{\text{m.m.f.}}$$

$$\text{flux} = 5\,000 \times 5 \times 10^{-6}\,\text{Wb}$$

$$= 25 \times 10^{-3}\,\text{Wb}$$

$$= 25\,\text{mWb}$$

PROBLEMS ON CHAPTER NINE

(1) The current in a 5 H choke is increased from 1 A to 3 A in 0·01 s. Calculate the average induced e.m.f.
   (2) The flux in a core of a 500 turn coil is 0·01 Wb when a current of 2 A flows. Calculate the inductance of the coil.

(3) Two coils are inductively coupled such that a current change of 10 A in 0·01 s in the first induces an average e.m.f. of 1 000 V across the second. Calculate the value of the mutual inductance.

(4) Calculate the average value of the coefficient of self inductance of a 500 turn coil if the core flux changes from 0·01 Wb to 0·015 Wb when the current changes by 0·2 A.

(5) Calculate the number of turns of a coil if the e.m.f. induced across one side is 500 V by a field changing at the rate of 500 mWb/s.

CHAPTER TEN

The Method of Dimensions

10.1 INTRODUCTION

As shown in earlier chapters and emphasised throughout, a system of units covering mechanics, heat, light and electricity is based on four fundamental concepts. Any quantity (with the possible exception of luminous flux, which takes into consideration the sensitivity of the human eye—*see* Chapter 6) may be expressed in terms of these basic concepts and a *dimensional analysis* carried out using the expressions thus obtained. The method is called the *method of dimensions*.

10.2 USES OF THE METHOD OF DIMENSIONS

The method of dimensions may be used in many ways. The three most important are (1) unit conversion, (2) checking equation accuracy, (3) predicting equation form.

Unit conversion is necessary whenever it is required to determine the relationship between units of the same quantity as they occur in different systems. Though one can use conversion tables they are not always readily available and conversion factors for every quantity are too numerous to memorise. Using the method of dimensions only the conversion factors of three quantities need be memorised (since the unit of time is common to all systems) and conversion from one system to any other may easily be effected. Unit conversion is demonstrated in Section 10.3.

The second use listed is important whenever it is necessary (a) to verify that the left-hand side of an equation is dimensionally the same as the right-hand side, for if it is not the equation is incorrect, and (b) to verify that a quantity which is expressed in terms of other and often more obscure quantities (in a dimensional sense) is dimensionally correct. This is clarified in Section 10.5.

Often in original work when it is necessary to obtain a defining equation of one kind or another it is possible to list the factors which may determine the equation and by comparison of indices of like quantities on either side of the equation the correct values may be obtained. The constants in such equations may then be determined by scientific observation. This is discussed in Section 10.6.

10.3   UNIT CONVERSION: ABSOLUTE SYSTEMS

Firstly the basic dimensions of the quantity are written down in terms of mass M, length L and time T. Then in place of each quantity the ratio of the basic unit of the system *from* which we are converting to that of the system *to* which we are converting is written in place of the basic symbol. The overall conversion factor is then computed.

Example 10.1
Determine the ratio of poundal to newton.

Both units are units of force in absolute systems. Force is equal to (mass) × (acceleration) and acceleration is expressed in units of length per unit time per unit time (*e.g.* m/s^2, ft/s^2, etc.). Hence force *dimensionally* is written

$$\frac{ML}{T^2} \qquad (10.1)$$

i.e.

$$\frac{(\text{mass}) \times (\text{length})}{(\text{time})^2}$$

Now

$$1 \text{ ft} = 0\cdot304\ 8 \text{ m}$$

$$1 \text{ lb} = 0\cdot453\ 6 \text{ kg}$$

We are converting *from* FPS *to* MKS (absolute).
The ratio of

$$\frac{\text{pound}}{\text{kilogramme}} = \frac{0\cdot453\ 6}{1}$$

which is inserted in the place of M in eqn. (10.1).
The ratio of

$$\frac{\text{foot}}{\text{metre}} = \frac{0\cdot304\ 8}{1}$$

which is written in the place of L in eqn. (10.1).

The conversion factor is then given by

$$\frac{\text{poundal}}{\text{newton}} = \frac{(0.453\ 6/1)(0.304\ 8/1)}{1^2}$$

(the second as basic time unit is common to both systems).
i.e.

$$\frac{\text{poundal}}{\text{newton}} = 0.138\ 2$$

i.e.

$$1 \text{ poundal} = 0.138\ 2 \text{ newton}$$

Example 10.2
How many joules are equal to one foot poundal?

Both these quantities are units of energy. Energy may be expressed as (force) × (distance) and from Example 10.1 force is written as ML/T^2 dimensionally. The unit of distance is, of course, the unit of length, written dimensionally as L. Hence, dimensionally energy is written as

$$\frac{ML}{T^2} \times L$$

i.e.

$$\frac{ML^2}{T^2} \tag{10.2}$$

We are again converting *from* FPS *to* MKS (since we wish to obtain an expression 1 ft pdl = ? joules), *i.e.* 1 FPS unit of energy = ? MKS units of energy.
Using the basic conversion factors from Example 10.1,

$$\frac{\text{foot}}{\text{metre}} = 0.304\ 8 \quad \text{and} \quad \frac{\text{pound}}{\text{kilogramme}} = 0.453\ 6$$

$$1 \text{ ft pdl} = (0.453\ 6)(0.304\ 8)^2 \text{ J}$$

$$= 0.042\ 1 \text{ J}$$

Example 10.3
Find the number of ft pdl/s in 1 W.

In this example we are converting the unit of power in the International System (MKS based) to that in the Imperial Absolute System (FPS based).

The conversion factors are

$$\frac{\text{metre}}{\text{foot}} = \left(\frac{1}{0\cdot304\ 8}\right) \quad \text{and} \quad \frac{\text{kilogramme}}{\text{pound}} = \left(\frac{1}{0\cdot453\ 6}\right)$$

and power (which is energy per unit time), written dimensionally, is

$$\frac{ML^2}{T^3} \qquad (10.3)$$

Hence

$$1\ W = \left(\frac{1}{0\cdot453\ 6}\right)\left(\frac{1}{0\cdot304\ 8}\right)^2 \text{ ft pdl/s}$$

$$= 23\cdot75 \text{ ft pdl/s}$$

This may easily be checked by reference to Example 10.2 in which the FPS energy unit was converted to MKS. The present example involves power and hence an additional power of T, but since the second is common to both systems, the conversion is essentially the reciprocal of that of Example 10.2 and consequently the conversion factor (23·75) is the reciprocal of 0·042 1.

10.4 UNIT CONVERSION BETWEEN ABSOLUTE AND GRAVITATIONAL SYSTEMS

The procedure is as already indicated with the additional point to be noted of what to do with the gravitational constant g. If converting from absolute to gravitational, g must be *divided* into the conversion factor; if converting from gravitational to absolute, g must be *multiplied* by the conversion factor. In all cases the value of g is that of the gravitational system involved.

Example 10.4
Determine the relationship between newtons and pounds force.

From eqn. (10.1) force is written dimensionally as ML/T^2; the conversion factors are

$$\frac{\text{metre}}{\text{foot}} = \left(\frac{1}{0\cdot304\ 8}\right) \quad \text{and} \quad \frac{\text{kilogramme}}{\text{pound}} = \left(\frac{1}{0\cdot453\ 6}\right)$$

and the value of g (since the gravitational system involved is the FPS-based system) will be taken as 32. Hence

$$1\ N = \left(\frac{1}{0\cdot304\ 8}\right)\left(\frac{1}{0\cdot453\ 6}\right)\left(\frac{1}{32}\right) \text{ lbf}$$

$$= 2\cdot26 \text{ lbf}$$

Had we been converting from FPS to MKS the factors would have been 0·304 8 for L, 0·453 6 for M; the value for g would be 32 as in the present example since the gravitational system involved is still the FPS.

Example 10.5
Determine the number of watts in one horsepower.

This example involves the conversion of FPS gravitational to MKS absolute and is interesting since a further constant is involved. It is first necessary to determine the number of watts in one foot pound force per second (since the horsepower is 550 ft lbf/s).
From eqn. (10.3) power is written dimensionally as

$$\frac{ML^2}{T^3}$$

We are converting from FPS to MKS (*i.e.* we require an equation 1 h.p. = ? W); the conversion factors are

$$\frac{\text{foot}}{\text{metre}} = 0\cdot304\,8 \quad \text{and} \quad \frac{\text{pound}}{\text{kilogramme}} = 0\cdot453\,6$$

The gravitational constant is 32 and must be multiplied. Hence

$$1 \text{ ft lbf/s} = 0\cdot453\,6 \times 0\cdot304\,8^2 \times 32 \text{ W}$$
$$= 1\cdot348 \text{ W}$$

Thus

$$1 \text{ h.p.} = 550 \text{ ft lbf/s} = 550 \times 1\cdot348 \text{ W}$$
$$= 741\cdot6 \text{ W}$$

The accepted figure is 1 h.p. = 746 W; the discrepancy is due to using 32 as an approximate figure for g. A more correct figure is actually 32·18 (but of course it varies from place to place which, as was pointed out in Chapter 3, is a disadvantage of a gravitational system).

10.5 CHECKING DIMENSIONAL ACCURACY OF EQUATIONS

The procedure is as follows:

(1) Write down the equation replacing the individual quantities by their dimensional equivalents in terms of M, L, T.

(2) Check powers of like *basic* quantities (M, L and T) on each side of the equation. If the powers of like quantities are not the same, the equation is dimensionally inaccurate. The method does not, of course, indicate the source of the inaccuracy, merely its presence.

Example 10.6
Check the accuracy of the equation

$$s = ut + \tfrac{1}{2}ft^2$$

where s is the distance moved in t seconds by a body having initial velocity u and an acceleration f.

Written dimensionally s (distance) is L; u (velocity) is L/T; f (acceleration) is L/T^2; t (time) is T. Hence

LHS is L

RHS is $\dfrac{L}{\not{T}} \times \not{T} + \tfrac{1}{2} \times \dfrac{L}{\not{T^2}} \times \not{T^2}$

Each term on the RHS has the dimension L, the term on the LHS has the dimension L, hence the equation is correct dimensionally. (Note that on the RHS the first term L is not added to the second term $\tfrac{1}{2}L$; such an exercise is meaningless.) For dimensional accuracy *each term* must have the *same* resultant dimension.

Example 10.7
The time τ taken for a capacitor to charge to approximately 64% of its final steady voltage is given by the equation

$$\tau = RC$$

where R is the resistance of the circuit in ohms, C is the capacitance of the capacitor in farads. (τ has a special name, the *time constant* of the circuit.) Check the dimensional accuracy of this equation.

This example requires a fourth symbol, one for the fourth basic concept electric charge, since the ohm and farad are electrical units. Q will be taken as the dimensional symbol for charge.

The ohm is the volt per ampere (*see* Chapter 7).
The volt is the joule per coulomb (*i.e.* energy/charge).
The ampere is the coulomb per second.

Hence the ohm is the volt/ampere, *i.e.* (joule/coulomb) × (second/coulomb), and from eqn. (10.2) the joule written dimensionally is ML^2/T^2, so that the volt written dimensionally is

$$\left(\frac{ML^2}{T^2Q}\right)$$

Hence dimensionally the ohm is

$$\left(\frac{ML^2}{T^2Q}\right) \times \left(\frac{T}{Q}\right)$$

i.e.

$$\frac{ML^2}{TQ^2} \tag{10.4}$$

The farad is the coulomb per volt and since the volt is the joule per coulomb the farad is the

$$\frac{\text{coulomb}}{\text{volt}} \quad i.e. \quad \frac{(\text{coulomb})^2}{\text{joule}}$$

written dimensionally as

$$\frac{Q^2T^2}{ML^2} \text{ [from eqn. (10.3)]} \tag{10.5}$$

Hence, multiplying the (ohm) × (farad), dimensionally we get

$$\frac{ML^2}{TQ^2} \times \frac{Q^2T^2}{ML^2} \quad i.e. \quad T$$

Since the dimension of the LHS is also T, the equation is dimensionally correct.

Example 10.8
The resonant frequency, f_r, of a circuit containing an inductor having inductance L henrys and a capacitor having capacitance C farads (the frequency at which the opposition to current offered by the inductor equals the opposition offered by the capacitor) is given by the equation

$$f_r = \frac{1}{2\pi(LC)^{\frac{1}{2}}}$$

Check the dimensional accuracy of this equation.

The special name given to the unit of inductance is the henry, which is defined by the equation (*see* Chapter 9)

back e.m.f. (volts) = inductance (henrys)
        × rate of change of current with time (amperes/second)

Hence a henry is actually a volt-second/ampere.

From Example 10.7 the volt is written dimensionally as ML^2/T^2Q, the ampere as Q/T, so that the henry may be written

$$\left(\frac{ML^2}{T^2Q}\right) T \times \frac{T}{Q} \quad i.e. \ \frac{ML^2}{Q^2}$$

From eqn. (10.5) the farad written dimensionally is

$$\frac{Q^2T^2}{ML^2}$$

so that the product (henry) × (farad)

$$= \frac{ML^2}{Q^2} \times \frac{Q^2T^2}{ML^2}$$

$$= T^2$$

The RHS of the equation which is to be checked contains the quantity

$$\frac{1}{[(\text{henry}) \times (\text{farad})]^{\frac{1}{2}}}$$

(the 2π, being a constant, has no bearing on the dimensional characteristic of the term).

Hence, dimensionally, the RHS is

$$\frac{1}{(T^2)^{\frac{1}{2}}} \quad i.e. \quad \frac{1}{T}$$

which is, of course, the dimension of frequency. (The unit of frequency, the hertz, is one cycle per second and since a 'cycle' cannot be expressed in terms of the concepts M, L, T and Q frequency is written $1/T$.)

10.6 PREDICTING EQUATION FORM

This use of the method of dimensions as described in Section 10.2 is illustrated by the following examples.

Example 10.9

The factors affecting the time for a complete oscillation of a simple pendulum (the periodic time) are the length of the pendulum, l, and the acceleration due to gravity, g; obtain the form of the equation connecting the periodic time, t, with these factors.

Suppose the powers to which l and g are raised are x and y respectively, so that the required equation is of the form

$$t = kl^x g^y \tag{10.6}$$

where k is a constant.

Writing this equation dimensionally, *i.e.* t as T, l as L and g as L/T^2 (acceleration), we obtain

$$T = kL^x \left(\frac{L}{T^2}\right)^y$$

$$= k \frac{L^x L^y}{T^{2y}}$$

$$= k \frac{L^{x+y}}{T^{2y}}$$

$$= kL^{x+y} T^{-2y}$$

Equating coefficients of L:
LHS the coefficient is zero (L does not appear, $L^0 = 1$). RHS the coefficient is $x + y$. Hence

$$x + y = 0 \tag{10.7}$$

Equating coefficients of T:
LHS the coefficient is 1. RHS the coefficient is $-2y$. Hence

$$1 = -2y$$

$$y = -\tfrac{1}{2} \tag{10.8}$$

and from eqn. (10.7)

$$x = +\tfrac{1}{2}$$

Substituting these values in the original equation, (10.6)—inserting the values in the dimensional equation leads to confusion

$$t = kl^{\frac{1}{2}} g^{-\frac{1}{2}}$$

$$= k \left(\frac{l}{g}\right)^{\frac{1}{2}}$$

k is determined by experimental observation (it is, in fact, equal to 2π).

Example 10.10

The kinetic energy of a moving body depends upon its mass and its velocity. Obtain the form of the relating equation.

Let x and y be the powers to which m and v are raised respectively. Then

$$\text{kinetic energy} = km^x v^y \qquad (10.9)$$

where k is a constant.

Dimensionally

$$\text{energy is written } \frac{ML^2}{T^2} \text{ [eqn. (10.2)]}$$

$$\text{mass is written } M$$

$$\text{velocity is written } \frac{L}{T}$$

so that

$$\frac{ML^2}{T^2} = kM^x \left(\frac{L}{T}\right)^y$$

Equating powers

$$\text{of } M \quad x = 1$$
$$\text{of } L \quad y = 2$$
$$\text{of } T \quad y = 2$$

Hence the equation [obtained by substituting these values in eqn. (10.9)] is of the form

$$\text{kinetic energy} = kmv^2$$

As before k is obtained by scientific experiment and observation (as was shown in Chapter 4, k is, in fact, equal to $\frac{1}{2}$ for an absolute system).

PROBLEMS ON CHAPTER TEN

(1) Using the method of dimensions *from first principles* express (a) one pound force in newtons, (b) one foot pound force in metre newtons. Take g as 32·18 ft/s².

(2) Check the dimensional accuracy of the following equations

$$\text{(a) } v^2 = u^2 + 2fs \qquad \text{(b) } s = ut + \tfrac{1}{2}ft^3$$

where u and v represent speed, f represents acceleration, s represents distance.

(3) Predict the form of the equation connecting the frequency f, at which the capacitive effect and inductive effect of a circuit containing a capacitor of C farads and an inductor of zero resistance and inductance L henrys are equal.

[*Hint:* The basic dimensions of frequency, inductance and capacitance are found in Chapters 6, 9 and 8 respectively. Write down the equation in the form $f_0 = kL^xC^y$ as shown in this chapter, then find x and y by equating powers of like quantities (k is a constant).]

(4) The period t of a mass m vibrating on a light spiral spring length l depends on the mass m, the restoring force of the spring when it is stretched F and the length l. Predict the equation form.

APPENDIX I

Standard Prefixes for Decade Multiples and Decimal Sub-Multiples of Units

Prefix	Abbreviation	Numerical Value
tera	T	10^{12}
giga	G	10^{9}
mega	M	10^{6}
kilo	k	10^{3}
hecto	h	10^{2}
deca	da	10^{1}
deci	d	10^{-1}
centi	c	10^{-2}
milli	m	10^{-3}
micro	μ	10^{-6}
nano	n	10^{-9}
pico	p	10^{-12}
femto	f	10^{-15}
atto	a	10^{-18}

1. Only one prefix should appear in any unit, *i.e.* pico rather than micromicro.

2. The symbol μ is commonly used to denote the millionth of the metre (10^{-3} mm) which is then termed a 'micron'.

APPENDIX II

The Relationship Between μ_0 and ε_0 and Its Implications

It was stated in Chapter 6 that electromagnetic radiation (heat, light, radiofrequencies, X-rays, etc.) is so called because it occurs whenever an electric field and magnetic field (Chapter 8) act together in a particular way. All electromagnetic radiation has a constant velocity of propagation (symbol c, value 3×10^8 m/s) often referred to as the 'velocity of light' (which it is, of course, but the fact that it is the velocity of all electromagnetic radiation must not be forgotten).

The electric field and magnetic field quantities ε_0 and μ_0, the permittivity of free space and the permeability of free space respectively, were discussed in detail in Chapter 8.

It is not surprising to learn that it has been quite conclusively shown that ε_0, μ_0 and c are inter-related. Exactly how will be stated shortly but not proved since the necessary field theory required is beyond the scope of this text. It can be shown that

$$ c^2 = \frac{1}{\mu_0 \varepsilon_0} $$

The equation is introduced to indicate the source of the old CGS Electromagnetic and Electrostatic Systems.

Before the relationship was known it was common practice to take μ_0 as unity when discussing electromagnetic quantities and ε_0 as unity when discussing electrostatic quantities. Now it is clear that if one of these quantities is unity the other cannot be. Thus this practice led to *two* separate systems of units, the CGS Electromagnetic System, in which $\mu_0 = 1$ (and ε_0 cannot be 1), and the CGS Electrostatic System, in which $\varepsilon_0 = 1$ (and μ_0 cannot be 1). To add to the general confusion another system, the practical system, was in existence (involving volts, amperes, etc.), and it was common for the research physicists and electrical scientists (the use of the word engineer is avoided) to express their findings in the 'academic' systems and for these to be applied many conversions, involving

155

powers of ten, had to be effected before the 'practical' advantage was gained. As we saw in Chapter 1, Professor Georgi tied in the 'practical' system with the MKS System to make *one* consistent system. Some engineers advocated the choice of 10^{-7} for the value of μ_0 (the power of ten chosen to give reasonable values of other derived quantities), but this choice leads to the circular constant 'π' appearing in relationships where no circular effect is involved and *not* appearing where one is. This system is the *MKS Unrationalised* System (broadly speaking one can interpret 'rational' as 'logical'). The SI is based on the *MKSA Rationalised* System in which $\mu_0 = 4\pi \times 10^{-7}$ (in which it follows from the equation relating ε_0, μ_0 and c that $\varepsilon_0 = 8 \cdot 85 \times 10^{-12}$) and the introduction of the '4π' removes the objection on logical grounds outlined above.

APPENDIX III

Comparison Between Systems

Term	Symbol	MKS	CGS	
			Electromagnetic	Electrostatic
Length	l	1 metre	100 cm	100 cm
Mass	m	1 kilogramme	1 000 g	1 000 g
Time	t	1 second	1 s	1 s
Force	F	1 newton	10^5 dynes	10^5 dynes
Work	W	1 joule	10^7 ergs	10^7 ergs
Power	P	1 watt	10^7 ergs/s	10^7 ergs/s
Current	I	1 ampere	0·1 unit	3×10^9 units
Charge	Q	1 coulomb	0·1 unit	3×10^9 units
e.m.f., p.d.	E, V	1 volt	10^8 units	1/300 unit
Resistance	R	1 ohm	10^9 units	$1·11 \times 10^{-12}$ unit
Inductance	L	1 henry	10^9 units	$1·11 \times 10^{-12}$ unit
Capacitance	C	1 farad	10^9 units	9×10^{11} units
Magnetomotive force	F	1 ampere-turn	$0·4\mu$ gilberts	—
Magnetising force	H	1 ampere-turn/metre	$4\pi \times 10^{-3}$ oersted	—
Magnetic flux	Φ	1 weber	10^8 maxwells	—
Magnetic flux density	B	1 weber/(metre)^2 (tesla)	10^4 g	—
Permeability of free space	μ	$4\pi \times 10^{-7}$ H/m	1 unit	—
Electric force	E	1 volt/metre	—	$1/3 \times 10^{-4}$ unit/cm
Electric flux	Ψ or Q	1 coulomb	—	$12\pi \times 10^9$ units
Electric flux density	D	1 coulomb/(metre)^2	—	$12\pi \times 10^5$ units/cm^2
Permittivity of free space	ε	$8·85 \times 10^{-12}$ F/m	—	1 unit

Answers To Problems

Chapter 2

(1) (a) 4·4 ft/s; (b) 3 mile/h. (2) (a) 30 mile/h; (b) 2 miles; (c) 2·2 ft/s².
(3) 2/3; 5/7 h. (4) (a) (i) 2·2 m/s², (ii) zero, (iii) 1 m/s²; (b) 45 s;
(c) No. (5) (a) 0·733 ft/s²; (b) 0·555 ft/s²; (c) 2·01 ft/s².

Chapter 3

(1) (a) 12·8 ft/s²; (b) 1·67 ft/s²; (c) 49 m/s²; (d) 12 m/s². (2) (a)
0·25 lbf, 9 pdl; (b) 10 N, 1·02 kp. (3) 28·8 ft/s². (4) (a) 4 lbf; (b) 24 lb;
(c) force = (mass) × (acceleration)/(g/6); (d) The value of g varies
considerably (down to zero as one leaves the Earth's gravitational
field). (5) 375 N. (6) (a) 1 kN; (b) 10·8 kN. (7) 4·6 ton f, 171·6 s.

Chapter 4

(1) 9 240 ft lbf, 0·28 h.p. (2) 40·7 s. (3) (a) 1·818 h.p.; (b) 960 000 ft
pdl; (c) 43·8 ft/s. (4) 1·96 kW. (5) 2·655 × 10⁶ ft lbf. (6) 14 m/s.

Chapter 5

(1) (a) 113°F; (b) 15°C; (c) 288·15 K; (d) 572·67°R. (2) (a) 10 050 cal;
(b) 116·5 J; (c) 0·753 6 Btu. (3) 50 J/kg K. (4) 35·17 s. (5) 0·559 7.

Chapter 6

(1) 3 150 lm. (2) 0·5 × 10¹⁵ Hz, red end of light band, 10 W/m².
(3) 5 443 foot-candles. (4) 24·6%. (5) 9 797·5 cd/m², 1 508 lx.

Chapter 7

(1) (a) 6·000 J; (b) 0·25 A/V; (c) 4 V/A; (d) 15 A. (2) A, 0·44 S;
B, 0·3 S; B. (3) 2·67 μWb/At. (4) This core, 100 000 At/Wb; core of
Problem 3, 375 000 At/Wb; this core. (5) 20 mC.

Chapter 8

(1) 2.12 A. (2) 398·25 pF; 0·2 μC. (3) 4 500 J. (4) 425 At. (5) 8 μF.

159

Chapter 9

(1) 1 000 V. (2) 10 H. (3) 1 H. (4) 12·5 H. (5) 1 000 turns.

Chapter 10

(1) (a) 1 lbf = 4·45 N; (b) 1 ft lbf = 1·356 5 mN. (2) (a) is dimensionally accurate, (b) the second term of the right-hand side has the wrong coefficient of t. (3) $f_0 = \mathrm{k}/(LC)^{\frac{1}{2}}$, where k is a constant. (4) $t = \mathrm{k}(ml/R)^{\frac{1}{2}}$, where k is a constant.

Index

Certain topics are referred to repeatedly throughout the book. In these cases the page numbers given are those where the main discussion or definition is presented.